The Physical Properties
of Liquid Metals

The Physical Properties
of Liquid Metals

TAKAMICHI IIDA
Osaka University, Japan

and

RODERICK I. L. GUTHRIE
McGill University, Montreal, Canada

CLARENDON PRESS · OXFORD
1988

Oxford University Press, Walton Street, Oxford OX2 6DP

Oxford New York Toronto
Delhi Bombay Calcutta Madras Karachi
Petaling Jaya Singapore Hong Kong Tokyo
Nairobi Dar es Salaam Cape Town
Melbourne Auckland
and associated companies in
Beirut Berlin Ibadan Nicosia

Oxford is a trade mark of Oxford University Press

Published in the United States
by Oxford University Press, New York

British Library Cataloguing in Publication Data
Iida, T.
The physical properties of liquid metals.
1. Liquid metals
I. Title II. Guthrie, Roderick I. L.
669'.9   TN690
ISBN 0–19–856331–0

Library of Congress Cataloging-in-Publication Data
I. Iida, T.
The physical properties of liquid metals
Bibliography: p.
Includes index.
1. Liquid metals. Guthrie, Roderick I. L. II. Title.
TN690.G894   1987   669   86–23616
ISBN 0–19–856331–0

Phototypeset by Macmillan India Ltd, Bangalore 25.
Printed by St Edmundsbury Press,
Bury St Edmunds, Suffolk

PREFACE

A knowledge of the physical chemistry of liquid metals (i.e. their physical properties and chemical behaviour) is required for a clear understanding of any liquid metal processing operations. During the last three decades, great progress has been made in the industrial technology and science of process metallurgy. The advance of science in extractive metallurgy has been described in a number of books such as *Physical chemistry of metals* by Darken and Gurry (1953), and *Physical chemistry of melts in metallurgy* by Richardson (1974). These books cover a wide range of topics relating to the physical chemistry of melts: structure, physical properties, thermodynamics, kinetic properties, and reaction kinetics. As such, they are of value, interest, and concern to process metallurgists. However, the former places emphasis on classical thermodynamics, while the latter is devoted mainly to classical thermodynamics and reaction kinetics. No book has yet laid emphasis on the structure and physical properties of metallic liquids for metallurgists and materials engineers, although the importance of these properties in liquid metal processing operations is well recognized. Although descriptions concerning the structure and physical properties of liquid metals appear to be largely ornamental in previous books, an understanding of relationships existing between a liquid metal's structure or physical properties, and various phenomena taking place during liquid metal processing operations, can be of significant help. This is especially true for the manufacture of high-quality metallic materials, where a detailed knowledge of the physical properties of the liquid metals or alloys in question can be of great help. Similarly, the atomistic or microscopic approach to process metallurgy, in contrast to equilibrium thermodynamics, can provide an essential framework to the understanding and interpretation of reaction rates in smelting and refining operations.

In present-day liquid metal processing operations, hydrodynamic analyses are playing an increasingly important role. For such continuum or macroscopic treatments, the values of physical properties are taken as empirical constants. If the atomistic approach proves to be successful, it will enable us to predict the values of these parameters. At the present time, however, accurate theoretical predictions are still not available, and experimental data for these constants are usually used. Unfortunately, considerable discrepancies often exist among experimental values for the physical properties of liquid metals. One would find it hard to judge which value is correct or most reliable when employed for hydrodynamic calculations, without some knowledge of the physical properties of liquid metals. As such, the subject is a necessary one for metallurgists and other engineers engaged in liquid metal processing operations.

v

As most books on continuum theory appear to be about thirty years out of date in their description of the physical properties of liquid metals, the object of this book is to introduce an outline of our present knowledge concerning theories, empirical relations, and experimental data on the physical properties of liquid metals. Furthermore, it is hoped that this book will be useful to readers with an interest in the atomistic approach to process metallurgy. This text is aimed at the graduate metallurgists and the materials engineers who deal with liquid metals. It has been assumed that the readers have some background knowledge in both elementary statistical mechanics and in elementary quantum mechanics. Theoretical equations are described, but detailed explanations of the theories are not given. The book is designed to provide a starting point for metallurgists or engineers with an interest in the atomistic nature of metallic liquids.

The book contains eight chapters and two appendices. General properties of liquid metals are given in Chapter 1. Chapter 2 is an introduction to the structure of liquid metals and alloys. Two fundamental quantities in the theory of liquids, the pair distribution function and the pair potential, are described. In particular, a detailed description of the pair distribution function is presented, because an understanding of the physical properties of liquids should be based on a fundamental understanding of their atomic arrangement. Chapter 3 is concerned with the density of liquid metals. Although density is an indispensable basic quantity, only a few review papers are available. Thermodynamic properties of liquid metals, i.e. vapour pressure, heat capacity, and sound velocity, are outlined in Chapter 4. Chapter 5 deals with the surface tension of liquid metals. Characteristic features of experimental data for liquid metal surface tensions are identified. Chapter 6 is devoted to the viscosity of liquid metals. Reasons for large discrepancies among experimental data for liquid metal viscosities are clarified in this chapter. Chapter 7 is concerned with diffusion in liquid metals; some of the problems of experimental investigations on diffusion measurements in liquid metals are discussed. In Chapter 8, the final chapter, electrical conductivity and thermal conductivity are outlined. The essential points of methods for measuring density, surface tension, viscosity, diffusivity, and thermal conductivity are also described. As a number of expressions presented in this book can be applied to molten salts, several examples are given in Appendix 1.

In conclusion, within the last three decades, great advances have been made in the theories of liquids. Similarly, a number of experimental studies have been carried out to help in our understanding of the behaviour and nature of liquid metals. However, our present knowledge of the physical properties of liquid metals is still not satisfactory from an engineering point of view. Accurate and reliable data for the physical properties of liquid metals and alloys are still not plentiful. Systematic investigations are greatly needed from the standpoint of basic engineering metallurgy.

Finally, we hope that the data on the physical properties of liquid metals given in this book are used not only as given constants, but will also help the user to recognize the importance of the microscopic approach to liquid metal processing operations.

Osaka and Montreal   T. I.
July 1986   R. I. L. G.

CONTENTS

PRINCIPAL SYMBOLS

Numbers in parentheses refer to equations.

CAPITAL ITALIC

A	area (e.g. molar surface; oscillating-plate)
A	a parameter (2.9); or a constant
A_i	area of component i (occupied in monolayer)
A^*	reduced area
B	a constant
B_S	isentropic bulk modulus
C	a constant
$C_{AW}(\eta)$	correction factor
C_P, C_V	heat capacity at constant pressure or volume
D	self-diffusivity; a constant
$D_i, D_{S,M}$	solute or solvent (or base metal) diffusivity
D^*	reduced rectilinear diameter (3.6); reduced diffusivity (7.24)
E	kinetic energy
E, E_a	resonant amplitude of plate in liquid or air (6.4)
E_F	Fermi energy
E_V^*	height of potential barrier
$\tilde{G}(Q)$	Fourier transform of $\{g(r)-1\}$
H	enthalpy
H	height
H^E	enthalpy of mixing (heat of mixing)
H_0	enthalpy of evaporation at $0\,K$
H_μ, H_D	apparent activation energy for viscous flow or diffusion
$1/H$	shape factor
$\Delta_s^l H_m$	enthalpy of melting
$\Delta_l^g H_b$	enthalpy of evaporation at T_b
$\Delta_s^g H_0$	enthalpy of sublimation at $0\,K$
I	intensity (of X-ray)
I	moment of inertia
I, I_0	intensity of emergent or incident X-ray beam (3.4)
K	coverage independent adsorption coefficient
K, K_0	apparatus constants
K_f	force constant
M	atomic weight (relative atomic mass); mass (6.9)

N	number of atoms
N_A	Avogadro's number
P	pressure
ΔP	pressure difference
$P(T)$	probability function
$P(A), P(B)$	proportion of vibrator A or B
P_1	Legendre polynomial
P_m	maximum bubble (gas) pressure
R	gas constant
R	radius
S	entropy
S^E	excess entropy
$\Delta_s^l S_m$	entropy of melting
$\Delta_l^g S_b$	entropy of evaporation at T_b
S_A	surface area
$S(Q)$	structure factor (interference function)
$S_{\alpha\beta}(Q)$	partial structure factor
T	absolute temperature
T	time period
T^*	reduced temperature ($T^* = T/T_c$; $T^* = Tk/\varepsilon$)
U	internal energy
U	velocity of sound
$U(Q)$	pseudopotential
U_c	cohesive energy
U_D	mobility
V	volume
V	atomic volume
V_A	atomic volume for a binary system
ΔV_m	volume change on melting
V^E	excess volume
$W(\phi)$	probability of a thermal fluctuation
$X, X', Y,$	
Z, Z'	parameters (for sessile drop method)
Z	first coordination number
Z	number of valence electrons (4.35), (4.36)
Z^E	excess valence
Z_i, Z_s	first coordination number within bulk or at surface

LOWER CASE ITALIC

a	average interatomic distance; a constant
a_s, a_i	activity of component s or i
b, c, d	constants

c	concentration
d	diameter of an atom (ionic radius after Pauling)
d	distance between solute and solvent ions
dv	volume element (2.4); increment of velocity (6.1)
e	electric charge
f	atomic scattering factor
f	surface-packing (configuration factor) (5.21)
f	activity coefficient (6.52)
$f(s), f_z$	interatomic forces
f, f_a	resonant frequency in liquid or air (6.4)
g	gravitational acceleration
$g(r)$	pair distribution function
$g_{\alpha\beta}(r)$	partial pair distribution function
h	depth of immersion
$h, \hbar$	Planck and Dirac constants ($\hbar = h/2\pi$)
i	chemical constant
k	Boltzmann's constant
k_F	radius of Fermi sphere (Fermi wave vector)
l	length
l	effective thickness of interface
m	mass of a single atom
m	mass of a drop
m, n	parameters
n	repulsive exponent
n	number of discrete ion (A.6)
n_0	average number density ($n_0 = N/V$)
$\dot{q}$	power input (8.5)
r	radial distance
r	radius
$r.d.f.$	radial distribution function
s	surface tension correction (3.1)
s	distance of displacement
t	time
u, v	pairwise interaction energy among atoms within bulk or on surface
v	volume of immersed suspension wire (3.1)
v	specific volume
v_0	close-packed molecular volume
v_F	Fermi velocity
v_f	average free volume per molecule
w	work or energy necessary to separate an atom
Δw	apparent loss of weight
x	atomic (or mole) fraction

x length
x, y, z coordinate axes

GREEK LETTERS

α isobaric thermal expansivity; absorption coefficient (3.4)

α_0 volume thermal expansion at 293 K

α, β phases

β a correction factor

β_s coefficient of sliding friction

Γ_s excess surface concentration

Γ_s^0 saturation coverage

γ surface tension

γ_M surface tension of mixtures

γ_0 surface tension of pure solvent (5.52); total molar surface energy

Δ dimensionless variable

δ, δ_0 logarithmic decrements

ε depth of attractive well; characteristic energy (ε/k, energy parameter)

ζ correction factor (for Roscoe's formula)

$\zeta_f, \zeta_H,$
ζ_s, ζ_{HS} friction coefficients

η packing fraction (packing density)

η_1 phase shift

θ contact angle; angle

θ Einstein characteristic frequency (6.21)

2θ scattering angle

θ_s fractional coverage (Γ_s/Γ_s^0)

κ transmission coefficient

κ_s isentropic compressibility

κ_T isothermal compressibility

Λ temperature dependence of liquid metal density

λ wavelength; screening radius (6.24)

λ thermal conductivity

μ viscosity; a parameter (bT_c/ρ_c)

μ_A viscosity of binary mixtures

μ_K, μ_ϕ viscosity due to kinetic contribution or pair interaction

μ^E excess viscosity

μ^* reduced viscosity

v, v_L mean frequency of atomic vibration or that which is calculated from Lindemann's formula

v kinematic viscosity

ξ, ξ_A parameters or constants

π	ratio of the circumference of a circle to its diameter (3.14159 . . .)
ρ	density
ρ_A	alloy density
$\Delta\rho$	fluctuation in number density
ρ_e	electrical resistivity
σ	effective hard-sphere diameter
σ_e	electrical conductivity
τ	shear force per unit area
Φ	total potential energy
ϕ	excess binding energy
$\phi(r)$	pair potential
$\tilde{\phi}_s(Q)$	Fourier transform of the long-range part of potential
χ	electronegativity
ω_t	transverse angular frequency

SUBSCRIPTS

b	at boiling point
c	at critical point
m	at melting point; first peak value of radial distance in $g(r)$ curve; maximum
g	gas
l	liquid
s	solid

ACKNOWLEDGEMENTS

This book came about thanks to the good offices of Professor Zen Morita, then Chairman of the Metallurgy Department, Osaka University, who first brought the two authors together and allowed for a year's sabbatical for T. I. at McGill University in 1981. Our approach to liquid metal properties bears the clear signature of his fine research activities and contributions to the subject, for which we are indebted. Similarly, the authors express their gratitude to their respective Universities, to the Natural Science and Engineering Research Council of Canada for financial support in the form of an International Scientific Exchange Award for T. I., and to Professor G. Farnell, then Dean of Engineering, for helping with a second period of support.

Miss K. Rivett, Mrs. P. Majumdar, and Miss J. Ritch deserve our warmest thanks for typing the various parts of the draft manuscripts, as do our families for being so understanding over the past six years.

The authors gratefully acknowledge permission to redraw, or reset, the following list of figures and tables from previous publications:

Figure/Table No.	Reference Source	Reprinted with permission from
Figure 1.1(a)	Giedt (1971)	Van Nostrand Company, New York, U.S.A.
Figure 1.2	Barber & Henderson (1981)	Scientific American: W. H. Freeman and Co., Publishers, New York, U.S.A.
Table 1.1	Tamamushi et al. (1981)	Iwanami Shoten, Publishers, Tokyo, Japan
	Kubaschewski & Alcock (1979a)	Pergamon Journals Ltd.
	Wilson (1965a)	The Metals Society, London, U.K.
Table 1.2	Kubaschewski & Alcock (1979b)	Pergamon Journals Ltd., Oxford, U.K.
Table 1.3	Wittenberg & DeWitt (1972a)	J. of Chemical Physics, New York, U.S.A.
Table 2.1	Waseda (1980d)	McGraw Hill, New York
Figure 2.23	Waseda (1980e)	McGraw Hill, New York
Figures 2.25, 2.26, 2.27, 2.28, 2.29	Kita, Zeze & Morita (1982)	Iron and Steel Inst. of Japan, Tokyo, Japan
Figure 3.3	Morita, Iida & Matsumoto (1985)	Japan Soc. for Promotion of Science, Tokyo, Japan
Figure 3.9, 3.10, 3.11	Dillon, Nelson, Swanson (1966)	J. of Chemical Physics, New York, U.S.A.
Figure 3.12	McGonigal (1962)	J. of Physical Chemistry, Washington, D.C., U.S.A.

Figure/Table No.	*Reference Source*	*Reprinted with permission from*
Figure 3.13	Steinberg (1974)	Metallurgical Transactions, Pittsburgh, Pa., U.S.A.
Figure 3.14	Iida & Morita (1978)	Japan Soc. for Promotion of Science, Tokyo, Japan
	Iida, Fukase & Morita (1981)	Japan Institute of Metals, Sendai, Japan
Figure 3.15	Iida & Morita (1978)	Japan Soc. for Promotion of Science, Tokyo, Japan
Figure 3.16	Predel & Emam (1969)	Materials Science & Engineering, Lausanne, Switzerland
Figure 3.19, 3.20, 3.21	Morita, Iida, Matsumoto (1985)	Japan Soc. for Promotion of Science Tokyo, Japan
Figure 4.1, 4.2(a), to 4.2(g), 4.3(a), 4.3(b) Tables 4.2, 4.3 and 4.4	Kubaschewski & Alcock	Pergamon Journals, Ltd., Oxford, U.K.
Figures 4.12 and 4.13	Gitis & Mikhailov (1967)	Soviet Physics Acoustics New York, U.S.A.
Figures 5.1	Inouye & Choh (1968)	Iron and Steel Institute of Japan, Tokyo
Figure 5.2	Morita, Iida & Kasama (1976)	Japan Institute of Metals, Sendai, Japan
Table 5.2	Egelstaff & Widom (1970)	Journal of Chemical Physics, New York, U.S.A.
Figures 5.14, 5.15	Halden & Kingery (1955)	Journal of Physical Chemistry, Washington, D.C.
Figure 5.16	Allen (1972d)	Marcel Dekker, New York, U.S.A.
Figure 5.17	Ogino, Nogi & Yamase (1980)	Tetsu-to-Hagane, Tokyo, Japan
Figure 5.18	Mori *et al.* (1975)	Japan Institute of Metals, Sendai, Japan
Figures 5.19, 5.20	Olsen & Johnson (1963)	Journal of Physical Chemistry, Washington, D.C., U.S.A.
Figures 5.21, 5.22, 5.23	Belton (1976)	Metallurgical Transactions, Pittsburgh, U.S.A.
Figures 5.24	Bernard & Lupis (1971)	Metallurgical Transactions, Pittsburgh, U.S.A.
Figure 6.6	Morita *et al.* (1984)	Tetsu-to-Hagane, Tokyo, Japan
Figure 6.7	Iida, Kijima & Morita (1983)	Tetsu-to-Hagane, Tokyo, Japan
Figures 6.8, 6.12, 6.13, 6.15, 6.16	Iida, Satoh, Ishiura Ishiguro & Morita, (1980)	Japan Institute of Metals, Sendai, Japan
Figure 6.14	Iida, Kumada, Washio & Morita (1980)	Japan Institute of Metals, Sendai, Japan

ACKNOWLEDGEMENTS

Figure/Table No.	Reference Source	Reprinted with permission from
Figures 6.17, 6.19	Iida & Morita (1980)	Japan Institute of Metals, Sendai, Japan
Figure 6.20	Pasternak (1972)	Journal of Physical Chemistry, Washington, D.C., U.S.A.
Figure 6.21	Iida, Guthrie & Morita (1982)	Canadian Institute of Metallurgy, Montreal, Canada
Figure 6.23	Iida, Morita, Takeuchi (1975)	Japan Institute of Metals, Sendai, Japan
Figure 6.24	Grosse (1961b)	Journal Inorg. Nuclear Chemistry, New York, U.S.A.
Figure 6.25	Iida et al. (1975)	Japan Institute of Metals, Sendai, Japan
Table 6.7	Iida et al. (1976)	Tetsu-to-Hagane, Tokyo, Japan
Figures 6.30, 6.31, 6.32, 6.33	Iida et al. (1976)	Tetsu-to-Hagane, Tokyo, Japan
Table 7.3	Protopapas, Anderson & Parlee	Journal of Chemical Physics, New York, U.S.A.
Table 7.4		
Figure 7.5	Pasternak (1972)	Journal of Physics and Chemistry of Liquids, Gordon and Breach, London, U.K.
Figures 7.8, 7.9, 7.10	Protopapas & Parlee (1976)	High Temperature Science, Clifton, New Jersey, U.S.A.
Table 7.4	Solar & Guthrie	Metallurgical Transactions.
Figure 7.11	(1972)	Pittsburgh, Pa, U.S.A.
Figure 7.12	Ma & Swalin (1960)	Acta Metallurgica, Bethesda, Md., U.S.A.
Figure 7.13	Ono (1977)	Tetsu-to-Hagane, Tokyo, Japan
Figure 7.14	Sacris & Parlee (1970)	Metallurgical Transactions, Pittsburgh, Pa., U.S.A.
Figure 8.1	Adams & Leach (1967)	Physical Reviews, New York, U.S.A.
Figure 8.2	Kita, Oguchi &	Tetsu-to-Hagane, Tokyo, Japan
Table 8.2	Morita (1978)	
Figure 8.3	Ono & Yagi (1972)	Iron and Steel Institute of Japan, Tokyo, Japan
Figure 8.4	Faber & Ziman (1965)	Philosophical Magazine, Univ. of Leicester, U.K.
Figure 8.5, 8.6, 8.7	Kita & Morita (1984)	North Holland Physics Publishing, Amsterdam, Holland
Figure 8.8	Cusack, Kendall & Fielder (1964)	Philosophical Magazine, Univ. of Leicester, U.K.
Figure 8.9	Duggin (1969)	Plenum Press, New York-London
Figure 8.12	Powell & Childs (1972)	McGraw Hill Book Co. New York-Toronto-London

1

GENERAL PROPERTIES

1.1. INTRODUCTION

1.1.1. Liquid region

It is well known that virtually all elements and chemical compounds can exist in solid, liquid, and gaseous forms depending on conditions of temperature and pressure. This is shown in the $P-V-T$, $P-T$, and $P-V$ phase diagrams of Figs. 1.1(a), 1.1(b), and 1.1(c), respectively. Metals exist in one of these forms.

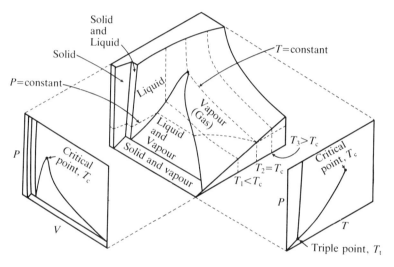

FIG. 1.1(a). $P-V-T$ phase diagram for a simple substance (after Giedt 1971).

At the triple point, the three states of matter—solid, liquid, and gas—are in equilibrium. The critical point represents the upper limit of the matter's possible existence as a liquid. Above this critical point no distinction between the liquid and the gas (vapour) can be made, and a single, undifferentiated, fluid state of uniform density exists. By contrast, however, there is no upper critical point for melting. Below the critical point, two first-order transitions are observed: melting and evaporation. The lower limit for the matter's existence as a liquid is generally taken to be its triple temperature T_t. However, the phenomenon of supercooling is well known and the liquid may exist in the supercooled state at $T < T_t$.

1

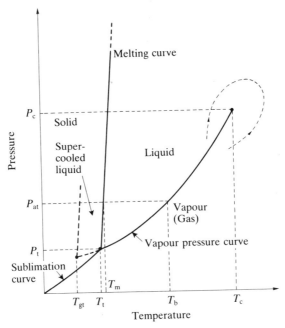

FIG. 1.1(b). *P–T* phase diagram for a simple substance. The temperatures of glass transition T_{gt}, triple T_t, melting T_m, boiling T_b, and critical points T_c, are indicated. The pressures of triple point P_t, atmosphere P_{at}, and critical point P_c are also indicated. The arrowed broken path represents a change from liquid to vapour (or from vapour to liquid) without phase transition. In the supercooled state, the liquid is metastable.

Liquid metals and alloys are ordinarily treated and examined from the scientific and technological point of view at atmospheric pressure. Table 1.1 lists values of melting temperature T_m, boiling temperature T_b, and liquid range for various metallic elements under normal conditions of atmospheric pressure (1.01325×10^5 Pa). In addition, Table 1.1 conveniently lists atomic weights (or relative atomic masses), since these values are required in the formulation from theory of some of the physical properties of the liquid metals. Examples include atomic volumes, Lindemann's melting law, viscosities, diffusivities, and sound velocities.

1.1.2. Characteristics of solids, liquids, and gases

It is first appropriate to consider the characteristic features of the liquid state and to compare its structure, from the microscopic standpoint, with those of the solid and gaseous states. As such, our discussion is general and not limited specifically to metallic liquids. During the past eight to ten decades, numerous studies have been carried out on the properties and characteristics of the solid

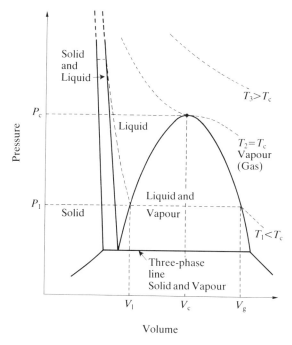

Pressure

P_c

P_1

Solid
and
Liquid →

$T_3 > T_c$

Liquid

$T_2 = T_c$
Vapour
(Gas)

Liquid and
Vapour

Solid

$T_1 < T_c$

Three-phase
line
Solid and Vapour

V_1 V_c V_g

Volume

FIG. 1.1(c). *P–V* phase diagram for a simple substance. V_1 and V_g are the volumes of the liquid and vapour coexisting in equilibrium at temperature T_1 and pressure P_1. V_c is the volume of the critical point.

and gaseous states. Both experimental and theoretical methods have been employed, and a considerable amount of knowledge has been accumulated. In contrast to that for the solid and gaseous states, our understanding of liquid structures and properties is still relatively imperfect despite at least equivalent efforts having been devoted to such investigations. A major reason for this difficulty is that the liquid state lacks any 'idealized model', on which to establish a base.

The concept of idealized models is very important in studying the properties of matter. Such a model can be defined as a hypothetical substance which has the characteristic features of real matter. Preferably, it should be relatively simple to treat mathematically, and its concepts should be clear. Such idealized models exist for the solid and gaseous states.

Thus, in an ideal solid or a perfect crystal, atoms are regularly arranged at the lattice points. This regular arrangement of atoms is long-range and three-dimensional, undisturbed by thermal agitation. The ideal solid forms a crystal of invariant shape.

By contrast, in the ideal gas, each atom can freely translate throughout the volume in which it is contained. These simple models lead to useful results for real solids and gases (e.g. Einstein's formula for specific heat, the Brillouin zone, and the Boyle–Charles Laws). Extensions and corrections to these idealized models for describing real substances can provide close agreement with experimental data.

Unfortunately, it is considerably more difficult to explain the behaviour of a liquid. From a microscopic point of view, the most characteristic feature of a liquid is its inability to support any shearing. This is manifest in its capacity to flow. Thus, the viscosities of liquids are very low, and diffusivities are very high, compared with equivalent properties in solids. On an atomic scale, this information suggests that an atom in the liquid state can easily migrate through fluctuations in density arising from the thermal motion of surrounding atoms. If we consider an atom in the liquid state at any moment, it will be interacting with the atoms which surround it, vibrating as though it were an atom in the solid state. At the next moment it may behave as an atom in the gaseous state, and move freely. It repeats such motions in changing from place to place. One sees that both atomic distance and time scales are of critical importance for a clear understanding of the structure and properties of liquids.

Figure 1.2 shows the trajectories of molecules in the solid, liquid and gaseous states, which were simulated with the aid of a computer (Barker and Henderson 1981). This figure indicates the characteristic features of the three states of matter.

Although it is often said that the liquid state exhibits intermediate properties between those of a solid and a gas, it should be noted that liquid properties cannot merely be averaged between those of their respective solid and gaseous states. Thus while matter is solid at low temperatures, gaseous at high temperatures and liquid at intermediate values (Fig. 1.1), one finds that a liquid will resemble a solid, in some ways, but in other ways it will more resemble a gas.

Liquids and solids exhibit similar densities. For example, in the case of metals, their densities in the solid and liquid states differ by less than 2–5 per cent near their melting points (Table 1.3). Both have similar number densities (i.e. number of atoms per unit volume) and average interatomic distances.

Furthermore, the ratio of the enthalpy of melting, $\Delta_s^l H_m$, to the enthalpy of evaporation, $\Delta_l^g H_b$, for metallic elements is of the order of 2–6 per cent with the exceptions of silicon, germanium and antimony (Table 1.2). Consequently, liquids and solids (the so-called condensed phases) resemble each other in their properties of cohesion.

At temperatures exceeding the critical point of the substance, as already mentioned, there is no discontinuous change from liquid to gas (vapour). In other words, the liquid and the gas can no longer be distinguished at the critical point: they have no permanent structure. From this fact, a liquid may then be considered as a dense gas. Both gases and liquids are often referred to as 'fluids'.

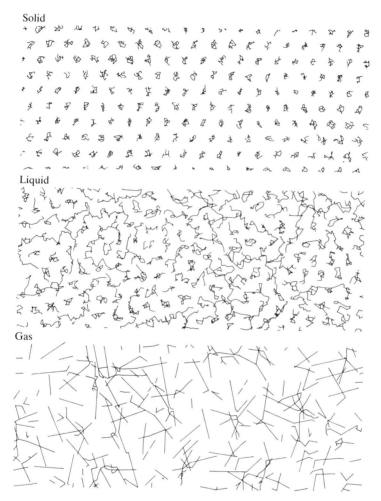

Solid

Liquid

Gas

FIG. 1.2. Trajectories of molecules in a solid, a liquid, and a gas. These were simulated with the aid of a computer. The two-dimensional system of molecules has the same phases and phase transitions as a real substance, but the molecular positions and motions are more easily displayed. In the solid the molecules are constrained to vibrate about fixed lattice sites, whereas molecules in the liquid and gaseous phases are free to wander. The only substantial differences between the two fluid states are those of density and of frequency of collision. The computer program calculates the trajectories by solving the equations of motion for some 500 two-dimensional molecules. The simulation was done by Farid Abraham of the International Business Machines Corporation Research Laboratory in San Jose, California (after Barker and Henderson 1981).

TABLE 1.1 *Values of atomic weight M, melting point T_m, boiling point T_b, and liquid range of metals and semi-metals at atmospheric pressure*

Metal, semi-metal		M	T_m (K)	T_b (K)	Liquid range, $T_b - T_m$ (K)
Lithium	Li	6.94	452.15	1590.15	1138
Beryllium	Be	9.01218	1558.15	3053.15	1495
Boron	B	10.81	2273.15–2773.15	(2823.15)	(50–550)
Carbon	C	12.011	–	–	–
Sodium	Na	22.98977	371.05 ± 0.05	1150.65	779.60 ± 0.05
Magnesium	Mg	24.305	923.15	1380.15	457
Aluminium	Al	26.98154	933.35	(2333.15)	(1400)
Silicon	Si	28.085	1687.15	2608.15	921
Phosphorus	P	30.97376	317.15 (white)[a] 866.15 (red)[a]	553[b]	236
Potassium	K	39.098	336.75 ± 0.05	1035.35	698.60 ± 0.05
Calcium	Ca	40.08	1124.15	1513.15	389
Scandium	Sc	44.9559	1812.15	(3000.15)	(1188)
Titanium	Ti	47.9	1998.15	3533.15	1535
Vanadium	V	50.9415	(1973.15)	3623.15[a]	(1650)
Chromium	Cr	51.996	2178.15	(2473.15)	(295)
Manganese	Mn	54.9380	1517.15	2368.15	851
Iron	Fe	55.84	1808.15	3003.15	1195
Cobalt	Co	58.9332	1765.15	3458.15	1693
Nickel	Ni	58.70	1728.15	3448.15	1720
Copper	Cu	63.54	1356.15	2903.15	1547
Zinc	Zn	65.38	692.62	1203.15	510.53
Gallium	Ga	69.72	302.930 ± 0.005	2573.15 ± 150	2270 ± 150
Germanium	Ge	72.5	1231.65	(2973.15)	(1742)
Arsenic	As	74.9216	1090.15 (36 atm)	889.15 (sublimating point)	
Selenium	Se	78.9	493.35	958.15	464.8
Rubidium	Rb	85.467	311.65	969.15	657.5
Strontium	Sr	87.62	1070.15	1912.15 ± 5	842 ± 5
Yttrium	Y	88.9059	1782.15	3200.15	1418
Zirconium	Zr	91.22	2130.15	4578.15[a]	2448
Niobium	Nb	92.9064	2740	3173.15	433.15
Molybdenum	Mo	95.94	2895.15 ± 10	5073.15	2178 ± 10
Technetium	Tc	{98}	2413.15	(4523.15)[a]	(2110)
Ruthenium	Ru	101.0	(2723.15)	(3973.15)	(1250)
Rhodium	Rh	102.9055	2239.15	(4233.15)	(1994)
Palladium	Pd	106.4	1828.15	2473.15	645
Silver	Ag	107.868	1233.65	2253.15	1019.5
Cadmium	Cd	112.41	594.05	1039.15 ± 2	445.1 ± 2
Indium	In	114.82	429.55	2373.15	1943.6
Tin	Sn	118.6	504.99	2548.15	2043.16
Antimony	Sb	121.7	903.65	1913.15 ± 8	1009.5 ± 8
Tellurium	Te	127.6	722.95	1663.15	940.2

TABLE 1.1. (*continued*)

Metal, semi-metal		M	T_{m} (K)	T_{b} (K)	Liquid range, $T_{\mathrm{b}} - T_{\mathrm{m}}$ (K)
Caesium	Cs	132.9054	301.65	1033.15	731.5
Barium	Ba	137.33	1263.15	1810.15	547
Lanthanum	La	138.905	1193.15	3742.15	2549
Cerium	Ce	140.12	1068.15	3741.15	2673
Praseodymium	Pr	140.9077	1208.15	3400.15	2192
Neodymium	Nd	144.2	1297.15	3300.15	2003
Promethium	Pm	{145}	(1320)[b]	(2700)[b]	(1380)
Samarium	Sm	150.4	1345.15	2173.15	828
Europium	Eu	151.96	1099.15	1712.15	613
Gadolinium	Gd	157.2	1585.15	(3273.15)	(1688)
Terbium	Tb	158.9254	1629.15	(3073.15)	(1444)
Dysprosium	Dy	162.5	1680.15	(2873.15)	(1193)
Holmium	Ho	164.9304	1734.15	(2873.15)	(1139)
Erbium	Er	167.2	1770.15	(3173.15)	(1403)
Thulium	Tm	168.9342	1818.15	2000.15	182
Ytterbium	Yb	173.0	1097.15	1700.15	603
Lutetium	Lu	174.96	1925.15	3600.15	1675
Hafnium	Hf	178.4	2480.15	> 3473.15	> 993
Tantalum	Ta	180.947	3123.15	> (4373.15)	> (1250)
Tungsten	W	183.8	3655.15	5828.15[a]	2173
Rhenium	Re	186.207	3440.15 ± 60	5873.15[a]	2433 ± 60
Osmium	Os	190.2	(2973.15)	(5773.15)	(2800)
Iridium	Ir	192.2	2727.15	> 5073.15	> 2346
Platinum	Pt	195.0	2047.15	4077.15	2030
Gold	Au	196.9665	1336.15	2983.15	1647
Mercury	Hg	200.5	234.28	629.73	395.45
Thallium	Tl	204.3	575.65 ± 0.1	1730.15 ± 10	1154.5 ± 10
Lead	Pb	207.2	600.55	2023.15	1422.6
Bismuth	Bi	208.9804	544.10 ± 0.05	1833.15 ± 5	1289 ± 5
Polonium	Po	{209}	527.15	1235.15	708
Astatine	At	{210}	–	–	–
Francium	Fr	{223}	–	–	–
Radium	Ra	226.0254	(973.15)	(1413.15)	(440)
Actinium	Ac	227.0278	1323.15	–	–
Thorium	Th	232.0381	2088.15	> 3273.15	> 1185
Protactinium	Pa	231.0359	–	–	–
Uranium	U	238.029	1405.15	4091.15	2686
Neptunium	Np	237.0482	921.15	–	–
Plutonium	Pu	{244}	912.65	3508.15	2595.5
Americum	Am	{243}	1123.15–1473.15	–	–
Curium	Cm	{247}	–	–	–
Berkelium	Bk	{247}	–	–	–
Californium	Cf	{251}	–	–	–
Einsteinium	Es	{252}	–	–	–

TABLE 1.1. (*continued*)

Metal, semi-metal	M	T_m (K)	T_b (K)	Liquid range, $T_b - T_m$ (K)
Fermium	Fm {257}	–	–	–
Mendelevium	Md {258}	–	–	–
Nobelium	No {259}	–	–	–
Lawrencium	Lf {260}	–	–	–

Data, except for those bearing the superscripts a or b, are taken from Tamamushi *et al.* (1981).
[a] Data from Kubaschewski and Alcock (1979a).
[b] Data from Wilson (1965a).
Data in braces { } denote the mass number of the isotope having the longest half-life.
Data in parentheses are unreliable.
Signs > denote 'above'.

TABLE 1.2. *Enthalpy and entropy changes associated with phase transitions of metallic elements, and enthalpies of evaporation of liquid metals at their melting points*

Metal	$\Delta_s^l H_m$ (kJ mol^{-1})	$\Delta_s^l S_m$ (J K^{-1} mol^{-1})	$\Delta_l^g H_b$ (kJ mol^{-1})	$\Delta_l^g S_b$ (J K^{-1} mol^{-1})	$\Delta_s^l H_m / \Delta_l^g H_b$ (%)	$\Delta_l^g H_m$ (kJ mol^{-1})
Li	2.93	6.45	148	92.7	2.0	157
Be	12.22	7.83	292	106	4.2	–
Na	2.64	7.12	99.2	85.9	2.7	107
Mg	8.8	9.54	128	93.9	6.9	133
Al	10.46	11.2	291	104	3.6	306
Si	50.6	30.0	383	108	13.2	392
K	2.389	7.11	79.1	75.8	3.0	87.6
Ca	8.4	7.55	151	85.9	5.6	158
Sc	(16.7)	(9.22)	–	–	–	(344)
Ti	14.6	7.51	426	120	3.4	438
V	16.7	7.68	–	–	–	485
Cr	20.9	9.81	342	116	6.1	351
Mn	14.6	9.62	220	94.3	6.6	246
Fe	13.77	7.61	340	109	4.1	357
Co	15.48	8.76	–	–	–	396
Ni	17.15	9.94	375	118	4.6	400
Cu	13.0	9.59	307	108	4.2	318
Zn	7.28	10.5	114	96.6	6.4	119
Ga	5.590	18.4	270	100	2.1	280
Ge	36.8	30.4	328	106	11.2	340
Se	5.9	12.0	–	–	–	–
Rb	2.197	7.06	75.7	78.8	2.9	85.2
Sr	10.0	9.61	154	93.1	6.5	160

TABLE 1.2. (*continued*)

Metal	$\Delta_s^l H_m$ (kJ mol^{-1})	$\Delta_s^l S_m$ (J K^{-1} mol^{-1})	$\Delta_l^g H_b$ (kJ mol^{-1})	$\Delta_l^g S_b$ (J K^{-1} mol^{-1})	$\Delta_s^l H_m/\Delta_l^g H_b$ (%)	$\Delta_s^g H_m$ (kJ mol^{-1})
Y	11.51	6.40	367	102	3.1	396
Zr	19.2	9.01	–	–	–	581
Nb	29.3	10.7	–	–	–	–
Mo	35.6	12.3	590	121	6.0	600
Pd	16.7	9.15	361	112	4.6	351
Ag	11.09	8.99	258	104	4.3	266
Cd	6.40	10.8	100	96.2	6.4	104
In	3.26	7.58	232	98.9	1.4	239
Sn	7.07	14.0	296	103	2.4	294
Sb	39.7	43.9	165	88.7	24.1	195
Cs	2.09	6.90	66.5	68.3	3.1	76.6
Ba	7.66	7.64	–	–	–	(178)
La	8.49	7.12	406	108	2.1	409
Ce	5.23	4.88	–	–	–	–
Pr	11.3	9.39	–	–	–	–
Nd	7.134	5.53	–	–	–	–
Sm	8.91	6.62	165	79.6	5.4	183
Hf	(24.06)	(9.61)	571	117	(4.2)	(571)
Ta	24.7	7.51	–	–	–	761
W	(35.1)	(9.53)	824	141	(4.3)	(823)
Re	33.5	9.70	–	–	–	711
Ir	(26.4)	(9.72)	612	130	(4.3)	(628)
Pt	22.2	10.9	469	(107)	4.7	504
Au	12.76	9.55	343	109	3.7	358
Hg	2.301	9.83	59.1	93.8	3.9	61.3
Tl	4.31	7.47	166	95.1	2.6	173
Pb	4.81	8.02	178	88.0	2.7	189
Bi	10.88	20.0	179	97.4	6.1	194
Th	13.93	6.89	–	–	–	544
U	12.6	8.97	417	94.6	3.0	–
Pu	2.80	3.07	344	93.1	0.8	336

Data from Kubaschewski and Alcock (1979*b*).

$$|\Delta_l^g H_m| \simeq |\Delta_s^g H_m| - |\Delta_s^l H_m|$$

Data in parentheses denote theoretical (or extrapolated) values.

As yet, we have no idealized liquid, largely because of the fundamental difficulty of describing the liquid state. However, quite recently, some properties of liquid metals have been explained quite skilfully and successfully on the basis of a relatively simple model, in which the atoms are treated as hard, inert spheres, i.e. the hard-sphere model.

1.2. GENERAL PROPERTIES OF LIQUID METALS

1.2.1. Richard's rule

The solid–liquid and the liquid–gas phase transitions (first-order transitions) are accompanied by changes in entropies, densities and volumes.

Richard's rule states that the entropy of melting, $\Delta_s^l S_m$ for metals has an average value of $8.8 \text{ J K}^{-1} \text{ mol}^{-1}$:

$$\Delta_s^l S_m = S_{\text{liquid}} - S_{\text{solid}} = \frac{\Delta_s^l H_m}{T_m} \approx 8.8 \text{ J K}^{-1} \text{ mol}^{-1} \tag{1.1}$$

or

$$\frac{\Delta_s^l H_m}{R T_m} \approx 1.06 \tag{1.2}$$

where $\Delta_s^l H_m$ ($= H_{\text{liquid}} - H_{\text{solid}}$) is the enthalpy of melting, sometimes called the enthalpy of fusion, (Table 1.2), and R is the gas constant.

$\Delta_s^l S_m$ is positive, since the enthalpy of a liquid is greater than that of a solid. Figure 1.3 shows the correlation between $\Delta_s^l H_m$ and T_m, i.e. Richard's rule, for a number of metallic elements. This figure indicates that most metals lie in the vicinity of $8.8 \text{ J K}^{-1} \text{ mol}^{-1}$, excluding the semi-metals (silicon, germanium,

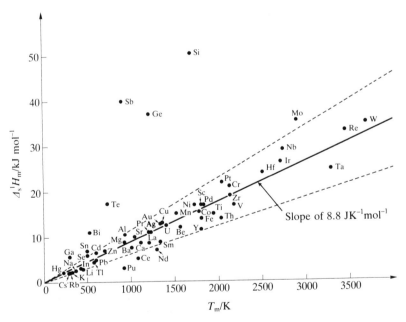

FIG. 1.3. Illustration of Richard's rule for various metallic elements. The slope of the continuous line is $8.8 \text{ J mol}^{-1} \text{K}^{-1}$. The broken lines denote ± 30 per cent error band.

tellurium, antimony, and bismuth). The dashed lines represent the ± 30 per cent error band.

It is known that the values of entropy and volume changes on melting depend somewhat on the crystal structures in the solid state. These are mentioned below.

1.2.2. Trouton's rule

Trouton's rule states that the entropy of evaporation, $\Delta_l^g S_b$, has an average value of 91.2 J K^{-1} mol^{-1}:

$$\Delta_l^g S_b = S_{gas} - S_{liquid} = \frac{H_{gas} - H_{liquid}}{T_b} = \frac{\Delta_l^g H_b}{T_b} \approx 91.2 \text{ J K}^{-1} \text{ mol}^{-1} \quad (1.3)$$

or

$$\frac{\Delta_l^g H_b}{RT_b} \approx 11.0 \quad (1.4)$$

where $\Delta_l^g H_b$ is the enthalpy of evaporation, sometimes called the enthalpy of vaporization, (Table 1.2).

Figure 1.4 indicates the correlation of $\Delta_l^g H_b$ with T_b for various metallic elements. Most metals lie within the ± 30 per cent error band. However, the metals with higher boiling points are apt to show larger deviations from Trouton's rule.

1.2.3. Relationship between enthalpy of evaporation (cohesive energy) and melting temperature

In discussing the physical properties of liquid metals, enthalpy of evaporation is one of their most important physical quantities. The term, enthalpy of evaporation, is used to denote that energy required to form separated, gaseous atoms from one mole of the liquid. The enthalpy of evaporation is a direct measure of the cohesive energy of a liquid metal. It is evident, therefore, that the physical properties of liquid metals (e.g. surface tension, thermal expansion, compressibility, and sound velocity), are related to their enthalpies of evaporation.

A liquid metal's melting temperature is also roughly related to its cohesive energy in a condensed state (i.e. solid or liquid). Indeed, the enthalpy of evaporation of liquid metals at their melting points, $\Delta_l^g H_m$, is approximately proportional to their melting temperature T_m: $\Delta_l^g H_m \simeq 2.3 \times 10^2 \, T_m$ (or $T_m \simeq 4.3 \times 10^{-3} \, \Delta_l^g H_m$), as shown in Fig. 1.5. To obtain values for $\Delta_l^g H_m$, the differences between the enthalpies of melting has been used: $|\Delta_l^g H_m| = |\Delta_s^g H_m| - |\Delta_s^l H_m|$. This approach is necessary since $\Delta_l^g H_m$ values for metallic elements at or near their melting points are not well established.

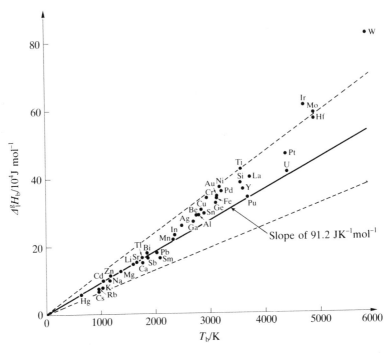

FIG. 1.4. Illustration of Trouton's rule for various metallic elements. The slope of the continuous line is 91.2 $J\,mol^{-1}K^{-1}$. The broken lines denote ± 30 per cent error band.

One should note from Fig. 1.5 that the Group II metals lie on their own straight line. In addition, several elements, in particular gallium, tin, and indium, show large deviations from this correlation. These elements have irregular or complex crystal structures. Thus, in solid gallium, there are pairs of atoms which are presumably bonded more tightly to each other than to other pairs. In melting, the pairs may be maintained, while in atomization they must be largely dissociated (Parish 1977).

Incidentally, the difference in the cohesive energy between a solid at $0\,K$, $\Delta_s^g H_0$, and the cohesive energies of liquid metals, $\Delta_l^g H_m$, is small, and for most metals the values lie within 10 per cent of each other (i.e. the average value of $\Delta_l^g H_m / \Delta_s^g H_0 \approx 0.94$) (Kittel 1971).

1.2.4 Volume change on melting

The majority of metals exhibit an average volume increase of 3.8 per cent during the solid–liquid phase transition. Volume contraction during melting

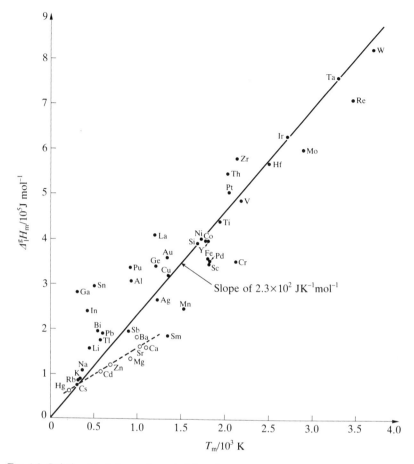

FIG. 1.5. Relationship between the enthalpies of evaporation and the melting temperatures for various metallic elements. The slope of the continuous line is $2.3 \times 10^2 \, \text{J mol}^{-1}\text{K}^{-1}$. ○ Group II metals.

occurs in the semi-metals gallium, silicon, germanium, and bismuth, plus cerium and plutonium. The volume increase is unusually small for antimony, lanthanum, praseodymium, neodymium, and neptunium.

On the basis of the volume change on melting, metallic elements can strictly be divided into two groups: Group 1, exhibiting volume increases during melting, and Group 2 characterized by volume decreases during melting. Most metals are in Group 1. According to Wittenberg and DeWitt (1972a) (Table 1.3), close-packed metals, i.e., face-centered cubic (f.c.c.) and hexagonal close-packed (h.c.p.), exhibit an average volume increase of approximately 4.6 per cent and an average entropy change of approximately $9.6 \, \text{J K}^{-1} \, \text{mol}^{-1}$. For

TABLE 1.3. *Volume and entropy changes associated with the melting of metallic elements*

Metal	Temperature (K)	$\Delta_s^l S_m$ (J K⁻¹ mol⁻¹)	$\Delta V_m/V_m$ (%)	Metal	Temperature (K)	$\Delta_s^l S_m$ (J K⁻¹ mol⁻¹)	$\Delta V_m/V_m$ (%)
Body-centred cubic structures (A2)[a]				Si	1683	29.8	−9.5
Li	454	6.61	2.74	Ge	1207	30.5	−5.1
Na	370	7.03	2.6	Bi	544	20.8	−3.87
K	337	6.95	2.54	Sb	904	22.0	(−0.95, +1.1)[b]
Rb	312	7.24	2.3				
Cs	302	6.95	2.6	Hg	234	9.79	3.64
Tl	575	7.07	2.2				
Fe	1809	7.61	3.6	*Lanthanide elements*			
				La	1193	5.61	0.6
				Ce	1073	4.85	−1.0
Close-packed structures (*f.c.c.* A1. *h.c.p.* A3)[a]				Pr	1208	5.73	0.02
Cu	1356	9.71	3.96	Nd	1297	5.48	0.9
Ag	1234	9.16	3.51	Sm	1345	6.40	3.6
Au	1336	9.25	5.5	Eu	1099	9.33	4.8
Al	931	11.6	6.9	Gd	1585	6.40	2.1
Pb	600	7.99	3.81	Tb	1629	6.53	3.1
Ni	1727	10.1	6.3	Dy	1680	8.54	4.9
Pd	1825	(9.6)	5.91	Ho	1734	9.41	7.5
Pt	2042	(9.6)	6.63				
In (A6)[a]	430	7.61	2.6	Er	1770	11.2	9.0
Mg	924	9.71	2.95	Tm	1818	9.67	6.9
Zn	692	10.7	4.08	Yb	1097	6.86	4.8
Cd	594	10.4	3.4	Lu	1925	7.15	3.6
Complex structures				*Actinide elements*			
Se	490	16.2	16.8	U	1406	8.62	2.2
Te	724	24.2	4.9	Np	913	5.69	1.5
Sn	505	13.8	2.4	Pu	913	3.18	−2.4
Ga	303	18.5	−2.9	Am	1449	9.92	2.3

(After Wittenberg and DeWitt 1972).
Data for volume and entropy changes on fusion are also given by Wilson (1965*b*), and by Ubbelohde (1965).
[a]See, for example, Barret (1966).
[b]Sb expands slightly on melting and has a small positive value for ΔV_m (Crawley, 1974).

the body-centered cubic (b.c.c.) metals, these values are somewhat smaller, the average volume increase being 2.7 per cent, and $\Delta_s^l S_m$ being 7.1 J K⁻¹ mol⁻¹. Although all the lanthanide metals, except erbium, thulium, and lutetium (which are f.c.c.), and the four actinide metals, have b.c.c. allotropes in equilibrium with the liquid, these elements were not included in the above averages.

The Group 2 metals which contract on fusion appear to be split into subgroups. The semi-metals (gallium, silicon, germanium, and bismuth) form the first, in which the $\Delta_s^l S_m$ values are much larger than those of the Group 1 elements. The other subgroup is composed of the elements cerium and plutonium, for which the $\Delta_s^l S_m$ values are smaller than for the Group 1 elements. Although the early lanthanide elements, lanthanum, praseodymium, and neodymium, do not contract during melting, their values for ΔV_m and $\Delta_s^l S_m$ are less than those in Group 1. The remaining elements in the lanthanide series appear to have nearly normal values. Similarly, in the actinide elements, the values for uranium are nearly normal, those for neptunium and plutonium are less than normal, and those for americium are, again, nearly normal.

The anomalous behaviour of the elements in Group 2 can be explained as follows (Wittenberg and DeWitt (1972). For the semi-metals, the rigid and directional bonds of the solid are apparently broken on melting, and the atoms become more spherical and pack closer together. On the other hand, in cerium, praseodymium, uranium, and plutonium their vacancies increase slightly, with corresponding reductions in their metallic radii.

1.2.5. Relationship between melting temperature and isobaric thermal coefficient of expansion (or expansivity)

It has long been known that metallic elements tend to follow a simple relationship between their melting temperatures T_m and their isobaric thermal

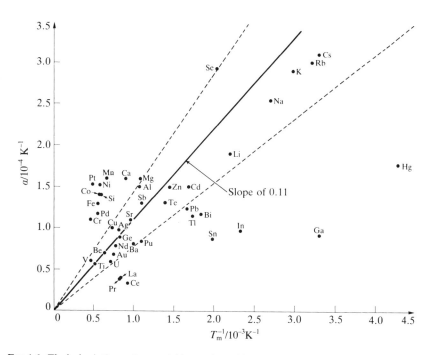

FIG. 1.6. The isobaric thermal expansivities vs. the melting points. The slope of the continuous line is 0.11. The broken lines denote ± 30 per cent error band.

expansivities $\alpha\{ \equiv V^{-1}(\partial V/\partial T)_p\}$, this being $\alpha T_m \approx 0.09$ (Turkdogan 1980; Steinberg 1974; see Table 3.1).

Figure 1.6 indicates this relationship. Apart from most transition metals, lathanide elements, and mercury, gallium, indium, and tin, about 23 metallic elements lie within the ± 30 per cent error band, shown by the dashed lines. However, this relationship is not particularly satisfactory as it has many exceptions.

Figure 1.7 shows a correlation between the isobaric thermal expansivity and the enthalpies of evaporation of liquid metals at their melting points (Table 1.2). As can be seen from Fig. 1.7, the correlation between α and $\Delta_l^g H_m$ is slightly better than that for α and T_m.

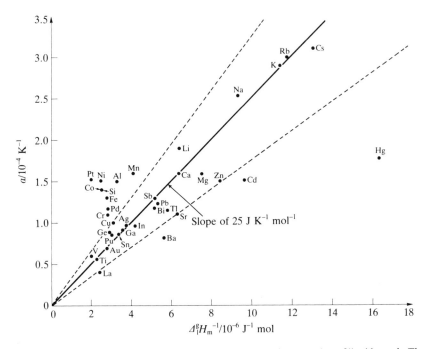

FIG. 1.7. The isobaric thermal expansivities vs. the enthalpies of evaporation of liquid metals. The slope of the continuous line is 25 J mol^{-1}K^{-1}. The broken lines denote ± 30 per cent error band.

1.2.6. Lindemann's equation (Lindemann's melting law)

Lindemann (1910) considered the melting process from the standpoint of the solid, and suggested that a solid shakes itself to pieces and changes into a liquid once the mean square displacement of each atom due to thermal vibration becomes a certain fraction (about 1/10) of the average interatomic distance.

The analysis results in

$$v = c\left(\frac{T_m}{MV^{2/3}}\right)^{\frac{1}{2}} \tag{1.5}$$

where v is the mean atomic frequency, M is the atomic weight, V is the atomic volume of the solid, and c is a constant which is roughly the same for all metals ($c \approx 9.0 \times 10^8$ in SI units).

Lindemann's formula was derived on the basis of dimensional considerations, so that values of c require experimental determination.

Singh and Sharma (1968), and Shapiro (1970) have examined Lindemann's melting law using the theory of lattice dynamics. Their results indicate that the value of c depends on the type of crystal lattice. If so, it follows that values of v evaluated on the basis of Lindemann's melting formula cannot be treated as being entirely reliable.

2

STRUCTURE

2.1. INTRODUCTION

In obtaining direct structural information on liquids, the results of x-ray and neutron diffraction experiments have proved to be of value. One finds that the x-ray diffraction pattern of the (perfect) crystalline solid consists of symmetrically sharp peaks. However, for the case of a gas, the x-ray diffraction patterns show a continuous scattering intensity with no maxima. This can be explained by a lack of any regular atomic arrangement in a gas at low density. According to experimental diffraction patterns for liquids which have been time and space averaged, liquids exhibit a few maxima and minima as shown in Fig. 2.1[†]. This can be interpreted as showing that atoms in the liquid state are randomly distributed in a nearly close-packed arrangement. Thus, one presumes that liquids have a certain amount of short-range order, which is a necessary consequence of high packing density, but have long-range disorder owing to thermal excitation and motion.

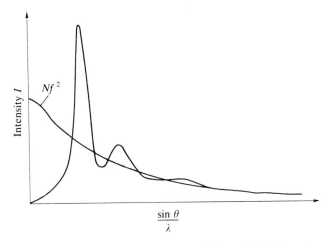

FIG. 2.1. X-ray diffraction pattern for a liquid (see e.g. Wagner 1972; Eisenstein and Gingrich 1942). Nf^2 curve corresponds to an (ideal) gas. I is the intensity of x-rays scattered through an angle 2θ, λ is the wavelength of the incident beam, N is the total number of atoms, and f is the atomic scattering factor.

[†] Even in the solid state, the diffraction patterns of imperfect solids which contain lattice defects exhibit peaks which become wider as the number of defects increases.

The structures of liquids closely resemble those of amorphous solids. The latter are generally obtained through rapid quenching from the liquid state. The amorphous solid provides a 'snapshot' of the atomic configuration of the liquid state. However, liquid structures averaged over time and over all atoms are slightly different from those of an amorphous solid, on account of significant differences in the degree of atomic motion between the two states.

2.2. DISTRIBUTION FUNCTIONS

2.2.1. Pair distribution function and radial distribution function

The description of liquids in terms of models has so far proved unsuccessful. However, in order to discuss the structure of liquids, it is first necessary to have a way of describing this structure mathematically. For this purpose, the pair distribution function $g(r)$ is of central importance to the modern theory of liquids, and the structures and properties of liquids in equilibrium are best described in terms of this function. However, it should be kept in mind that $g(r)$ is averaged over time and over all atoms, and does not give a 'snapshot' of an atomic distribution at any particular point in time.

Consider, therefore, a monatomic liquid in equilibrium, and then any atom with its centre at the point $r = 0$. The number of atoms, dN, in the spherical shell between radial distances r and $r + dr$ from the origin atom is

$$dN = 4\pi r^2 \, dr \left(\frac{N}{V}\right) g(r) \tag{2.1}$$

where N is the total number of atoms in the volume V. This pair distribution function depends on the magnitude of $\mathbf{r}$ but not its direction, since real liquids are generally isotropic.

If there is no correlation between atoms, as is the case for the ideal gas which consists of an identical mass of point atoms, $g(r)$ is everywhere equal to unity (i.e. $g(r) \neq f(r)$). Consequently, in the case of the ideal gas, the number of atoms in the spherical shell between r and $r + dr$ is

$$dN = 4\pi r^2 \, dr \, n_0 \tag{2.2}$$

where $n_0 (= N/V)$ is the average number density.

We now consider the pair distribution function in real liquids. The pair distribution function curve as a function of distance becomes complicated on account of the existence of atomic interactions, i.e. the forces of attraction or repulsion between atoms in the condensed state. For larger r, as one might expect, the value for $g(r)$ tends towards unity, because the correlations between the central and outlying atoms weaken rapidly with distance. In other words, at large r, the probability of finding another atom in the spherical shell between r and $r + dr$ is independent of the presence of the reference atom. This corresponds to complete disorder. On the other hand, for small values of r, that is, for values of r less than the atomic diameter, the probability must tend to zero, because two atoms cannot overlap.

Consequently, the pair distribution function $g(r)$ may be defined as follows: $g(r)$ is proportional to the probability of finding another atom, at the same instant, at a distance r from the reference atom located at $r = 0$†. In this respect, the function $4\pi r^2 n_0 g(r)$ is generally called the radial distribution function (r.d.f.).

A typical curve for the pair distribution function $g(r)$ is given in Fig. 2.2. As can be seen from this figure, there are a few maxima and minima in $g(r)$, and their amplitudes decrease rapidly towards unity with increasing r. This departure of $g(r)$ from unity shows the existence of short-range order around the reference atom. A pronounced first peak in $g(r)$ is located roughly on the minimum of the pair potential (see Fig. 2.5(c) and (d)).

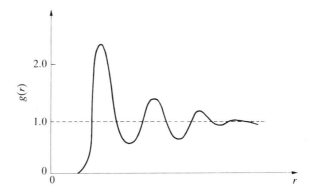

FIG. 2.2. Typical pair distribution function for a simple liquid.

A typical form of the r.d.f. is shown in Fig. 2.3. Since the pair distribution function approaches unity for large r, the radial distribution function draws close to the parabolic curve of $4\pi r^2 n_0$. The hatched area, or the area under the principal peak, can be interpreted as the number of nearest-neighbour atoms and the so-called first coordination number. The first coordination number Z can be expressed mathematically by

$$Z = 2 \int_{r_0}^{r_m} 4\pi r^2 n_0 \, g(r) \, dr \tag{2.3}$$

† The pair distribution function does not consider interactions between the pair of atoms and other atoms surrounding them.

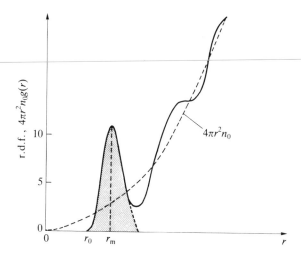

F<small>IG</small>. 2.3. Typical curve for the radial distribution function (r.d.f.) (see e.g. Ocken and Wagner 1966; Eisenstein and Gingrich 1942). r_0 and r_m represent the beginning and first peak values of radial distance in the r.d.f. curve, respectively.

Since the properties of liquids can be described approximately in terms of the first coordination number, it is frequently used as an important parameter. At the present time, however, we have no absolute method for estimating the first coordination number. It is important to note that its values may vary depending on the methods used in its estimation (e.g. Waseda 1980a).

2.2.2. Theoretical calculation of $g(r)$

According to statistical mechanical theory, the pair distribution function is described through the formula:

$$g(r) = V^2 \frac{\overset{(N-2)}{\int \cdots \int} e^{-\Phi/kT} dv_3 \ldots dv_N}{\underset{(N)}{\int \cdots \int} e^{-\Phi/kT} dv_1 \ldots dv_N} \qquad (2.4)$$

where V represents the total volume of the system, dv represents a volume element, and where Φ is the total potential energy of the system. The pair distribution function in the equation is expressed in terms of the total interatomic potential energy Φ. However, it is impossible, in practice, to calculate the pair distribution function by direct solution of this multi-integral equation. Consequently, several approximate equations, e.g. Born–Green[†], Percus–Yevick, and Hypernetted Chain equations, have been suggested (e.g. Egelstaff 1967a). These equations give us integral formulae connecting the pair distribution function $g(r)$ and the pair potential $\phi(r)$[‡]. In practice, these equations are sometimes used in reverse so as to calculate $\phi(r)$ from $g(r)$, which can be measured experimentally.

The theoretical calculation of the pair distribution function has proved to be a difficult task because of the mathematical complexity and lack of available information regarding the pair potential ϕ_{ij}. As yet, in the theoretical calculation of $g(r)$ there are no satisfactory results for engineering applications.

2.2.3. Experimental determination of $g(r)$

The experimental investigation of liquid structures can be carried out through the use of x-ray, neutron, or electron diffraction techniques.

The pair distribution function is obtained from measured intensities (see Fig. 2.4) by using the formula:

$$g(r) = 1 + \frac{1}{2\pi^2 n_0 r} \int_0^\infty Q\left(\frac{I}{Nf^2} - 1\right) \sin(Qr)\, dQ \qquad (2.5)$$

$$Q = \frac{4\pi \sin\theta}{\lambda}$$

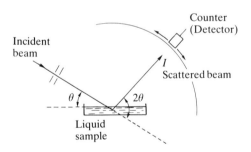

FIG. 2.4. Schematic representation of a scattering experiment.

where 2θ is the scattering angle (of x-rays), λ is the wavelength of the incident beam, f is the atomic scattering factor, i.e. the Fourier transform of the electron density in the atom, and I is the intensity of reflected beams from the liquid.

[†] This is sometimes called the Yvon–Born–Green (YBG) equation.
[‡] It is assumed that the total potential Φ can be decomposed into a sum of values of pair potential ϕ_{ij} (the potential developed between pairs of atoms i, j), such that $\Phi(r_1 \ldots r_N) = \Sigma \phi(r_{ij})$ for $i < j$.

This formula may also be interpreted as the definition of $g(r)$. The determination of $g(r)$ from observed intensities is subject to some problems. These stem mainly from experimental difficulties and from the methods used to treat the experimental data. The relative intensity of (I/Nf^2) is defined as a (liquid) structure factor or an interference function (denoted by $S(Q)$ in this book), which is frequently used in discussions concerning the structures of non-crystalline materials. For isotropic systems, $S(\mathbf{Q}) \equiv S(Q)$.

2.2.4. The pair potential

We may now introduce the pair potential $\phi(r)$. This represents the potential energy between an atom and its surrounding neighbours. All equilibrium properties of liquids can be expressed directly by using the pair distribution function $g(r)$ in conjunction with $\phi(r)$. For this reason, $g(r)$ and $\phi(r)$ can be regarded as being the most fundamental of liquid properties. Pair potentials $\phi(r)$ should be derivable, in theory, from quantum mechanics or electron theory. However, strictly speaking at the present time, quantum mechanical calculations of pair potentials are practically impossible apart from those of simple atoms such as hydrogen and helium. Consequently, various model pair potentials are frequently used as an alternative in numerical calculations, as follows.

2.2.4.1. Model potentials

(a) The hard-sphere potential (see Fig. 2.5(a)) is

$$\phi(r) = +\infty \quad \text{for } r < \sigma$$
$$\phi(r) = 0 \quad \text{for } r \geqslant \sigma \tag{2.6}$$

where σ is the diameter of the hard-sphere. Attractive forces are entirely neglected in this model.

(b) The inverse power potential (see Fig. 2.5(b)) is

$$\phi(r) = \varepsilon \left(\frac{\sigma}{r} \right)^n \tag{2.7}$$

where ε, σ and n are parameters. If n tends to infinity, the inverse power potential becomes that of the hard sphere.

2.2.4.2. The Lennard–Jones potential

This is used for insulating liquids (see Fig. 2.5(c)):

$$\phi(r) = 4\varepsilon \left[\left(\frac{\sigma}{r} \right)^m - \left(\frac{\sigma}{r} \right)^n \right] \tag{2.8}$$

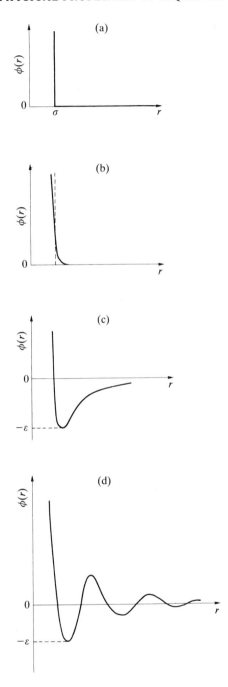

FIG. 2.5. Schematic representation of the pair potentials. ε is the depth of the attractive well.

The parameters m and n are generally taken as 12 and 6, although other combinations are frequently used. The Lennard–Jones (LJ) potential can be considered as a simple empirical expression that more recently has been shown to derive from quantum mechanics. It is useful for simple insulating liquids such as argon.

2.2.4.3. The effective ion–ion potential

This is used for liquid metals (see Fig. 2.5(d)). In the case of liquid metals, the pair potential corresponds to the effective ion–ion potential given in eqn (2.9) below:

$$\phi(r) = \frac{A}{r^3} \cos(2k_F r) \tag{2.9}$$

where A is a parameter, and k_F is the radius of the Fermi sphere, or the Fermi wave vector. The pair potentials for metals are assumed to be for long-range oscillatory interactions (Friedel oscillations) resulting from the presence of conduction electrons (Friedel 1952; Harrison 1966). However, details of the long-range oscillatory potential are not yet clear.

Parameters appearing in eqns (2.6) to (2.9) can be determined by experiments involving x-ray and/or neutron scattering (Johnson, Hutchinson, and March 1964; Waseda 1980b).

2.2.5. Some equations in terms of $g(r)$ and $\phi(r)$

In order to facilitate an understanding of the concept of the pair distribution function $g(r)$, let us derive the equations for equilibrium properties.

2.2.5.1. Total internal energy

For a monatomic liquid at temperature T, the total internal energy U_1 is given by

$$U_1 = \frac{3}{2} NkT + \langle \Phi \rangle \tag{2.10}$$

where k is Boltzmann's constant, and $\langle \Phi \rangle$ the mean total potential energy.

The term $(3\,NkT/2)$ in eqn (2.10) represents the liquid's total kinetic energy. Let us derive $\langle \Phi \rangle$.

(a) As mentioned previously, the average number of atoms within a radial distance between r and $r+dr$ of a given reference atom is $n_0 g(r)$ $(4\pi r^2\,dr)$.

(b) The average potential energy of interaction with these neighbours is $n_0\, g(r)\, \phi(r)\, (4\pi r^2\,dr)$.

(c) Integrating over r to obtain the total potential energy, and dividing by 2 to avoid counting each pair interaction twice over, results in

$$\langle \Phi \rangle = \frac{n_0 N}{2} \int_0^\infty g(r)\,\phi(r)\,4\pi r^2\,dr$$

Consequently, we obtain

$$U_1 = \frac{3}{2} NkT + \frac{n_0 N}{2} \int_0^\infty g(r)\,\phi(r)\,4\pi r^2\,dr \tag{2.11a}$$

or

$$U_1 = \frac{3}{2} NkT + \frac{2\pi N^2}{V} \int_0^\infty g(r)\,\phi(r)\,r^2\,dr \tag{2.11b}$$

Although eqn (2.11) has been derived through descriptive arguments, it is also possible to arrive at it through the partition function in statistical mechanics (e.g. Egelstaff 1976b).

2.2.5.2. Heat capacity at constant volume c_V

Combining eqn (2.11) with the thermodynamic formula for heat capacity at constant volume C_V immediately gives

$$C_V = \left(\frac{\partial U}{\partial T} \right)_V = \frac{3}{2} Nk + \frac{2\pi N^2}{V} \int_0^\infty \left[\frac{\partial g(r)}{\partial T} \right]_V \phi(r)\,r^2\,dr \tag{2.12}$$

Since the assumption is made that $\phi(r)$ is independent of temperature, only information on the variation of $g(r)$ with T at constant volume is required.

2.2.5.3. Enthalpy of vaporization $\Delta_1^g H$

The total internal energy of a gas or vapour, U_g, is $3NkT/2$, while the work of expansion against an external pressure during vaporization is NkT ($PV = NkT$) if the vapour behaves as an ideal gas. Consequently, the liquid's enthalpy of vaporization, or the heat of vaporization $\Delta_1^g H$, is given by

$$\Delta_1^g H = U_g - U_1 + NkT \tag{2.13}$$

$$= NkT - \frac{2\pi N^2}{V} \int_0^\infty g(r)\,\phi(r) r^2\,dr$$

2.3. THE STRUCTURE OF PURE LIQUID METALS

In Subsection 2.2.4, a theoretical analysis by Friedel predicted that the pair potential in a metal is of an oscillatory nature which becomes damped out at large distances. The form of this pair potential differs from that of a non-metallic liquid (see. Figs. 2.5(c) and 2.5(d)). Consequently, it is also expected that the pair distribution functions should show clear differences between the metallic and non-metallic liquids, since $g(r)$ is given as a function of $\phi(r)$. However, pair distribution functions obtained from experimental results show that the shape of $g(r)$ for liquid metals closely resembles non-metallic liquids such as argon.

A considerable amount of structural information for liquid metals has been accumulated during the past two decades. The values of pair distribution functions obtained from x-ray scattering experiments are shown in Figs. 2.6–2.9 for sodium, lead, copper, and iron near their melting points. All of the distribution curves have the same general shape; namely, the values of $g(r)$ show deviations from unity at distances of approximately one (a pronounced peak), two, three and four atomic diameters, and become equal to about unity beyond four atomic distances. These figures also show that the value of $g(r)$ at distances of approximately one-and-a-half atomic diameters (i.e. $r \approx 1.5\sigma$) are very small. The deviations in the curves from unity imply that short-range order holds to three or four atomic diameters in liquid metals.

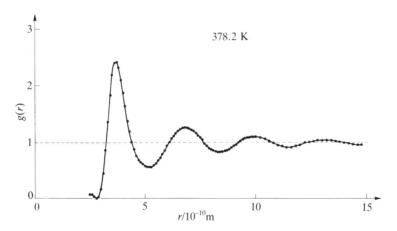

FIG. 2.6. Pair distribution function $g(r)$ of liquid sodium near the melting point.

Structure factors for these metals are shown in Figs. 2.10–2.13. Many workers have investigated the structure of liquid metals from the standpoint of the hard-sphere model (e.g. Furukawa 1960; Aschroft and Lekner 1966). They found that the hard-sphere potential, though a crude approximation, could,

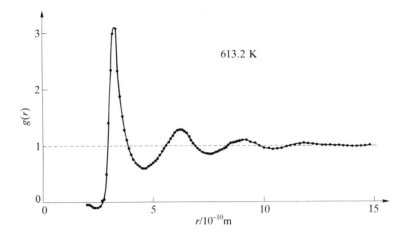

FIG. 2.7. Pair distribution function $g(r)$ of liquid lead near the melting point.

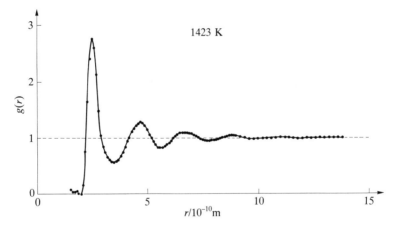

FIG. 2.8. Pair distribution function $g(r)$ of liquid copper near the melting point.

through suitable choice of hard-sphere diameter and packing fraction[†], reproduce measured structure factors. The results imply that the characteristic

---

[†] This is sometimes called the packing density. The packing fraction or the packing density η is defined by

$$\eta = \frac{4}{3}\pi\left(\frac{\sigma}{2}\right)^3\frac{N}{V} = \frac{\pi}{6}n_0\sigma^3$$

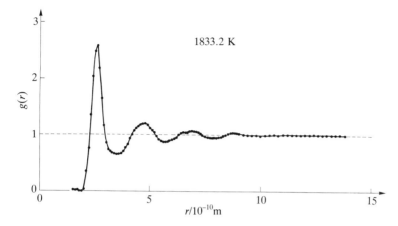

FIG. 2.9. Pair distribution function $g(r)$ of liquid iron near the melting point.

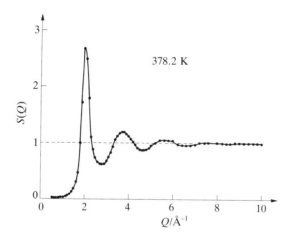

FIG. 2.10. Structure factor $S(Q)$ of liquid sodium near the melting point.

features of the structure factor of liquid metals are largely determined by ion–ion repulsions, and that the effect of the long-range oscillations in the pair potential on the structure factor is small for liquid metals and alloys. In practice, even though liquid metals have very complex interatomic interactions, these simplifications can be made.

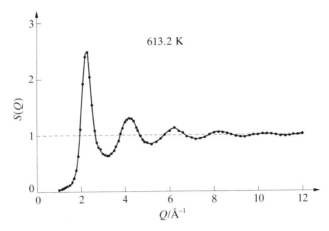

FIG. 2.11. Structure factor $S(Q)$ of liquid lead near the melting point.

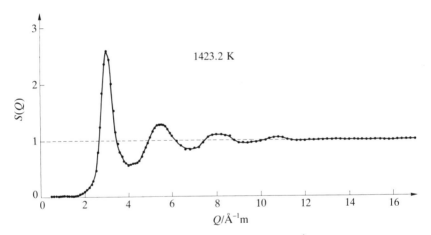

FIG. 2.12. Structure factor $S(Q)$ of liquid copper near the melting point.

As further support, one may note that the random packing of hard spheres also leads to reasonably good agreement with experimental data for the pair distribution function.

The best value for the packing fraction that is compatible with experimental data for the $S(Q)$ curves is about 0.45 for the majority of liquid metals near their melting points (Waseda 1980c).

The pair distribution functions obtained from scattering data for mercury and tin near their melting points are shown in Figs. 2.14 and 2.15. These curves of $g(r)$ are somewhat different from those of the simple metals such as sodium

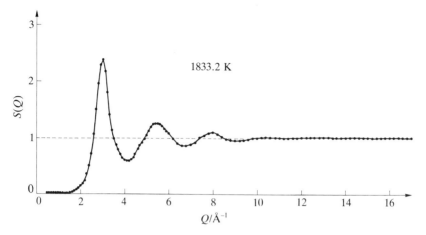

FIG. 2.13. Structure factor $S(Q)$ of liquid iron near the melting point.

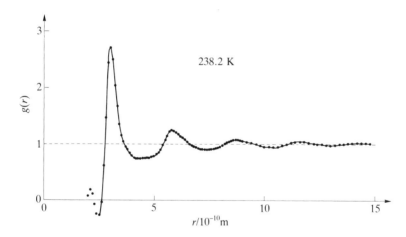

FIG. 2.14. Pair distribution function $g(r)$ of liquid mercury near the melting point.

and lead; namely, in the case of mercury there seems to be a slight asymmetry in the first peak. To this type of $g(r)$ curve belong indium, thallium, gadolinium and terbium. On the other hand, in the case of the $g(r)$ curves for tin compared with those for lead, there is a small hump or shoulder to the right of the first peak, as the large-scale graph in Fig. 2.16 shows. To this type belong tin, gallium, silicon, germanium, antimony, and bismuth. These metals also exhibit a small hump to the right of the main peak in $S(Q)$ (Waseda 1980c).

As mentioned previously, the structure of liquid metals can be described by the random packing of hard spheres. In a stricter sense, however, atoms in the liquid metals just mentioned can retain slightly-directional bonding, charac-

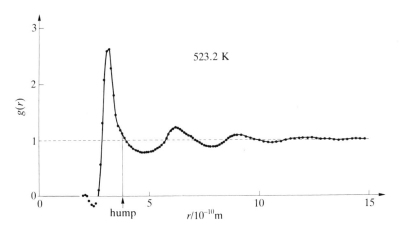

FIG. 2.15. Pair distribution function $g(r)$ of liquid tin near the melting point.

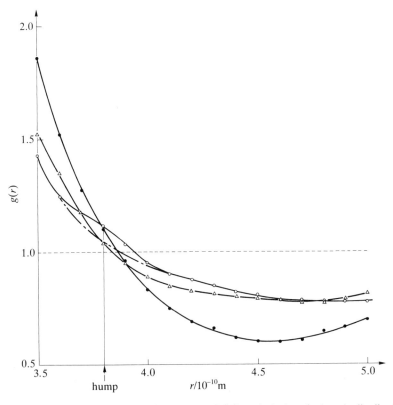

FIG. 2.16. Plots of the right-hand-side component of the principal peaks in pair distribution functions $g(r)$ for liquid tin and lead. ○ tin (523 K), △ tin (1373 K), ● lead (613 K).

teristic of their solid crystalline structures, despite thermal agitation in the liquid state.

As an example of the temperature dependence of the pair distribution function, values for liquid aluminium and tin are shown in Figs. 2.17 and 2.18.

In discussing the properties of liquid metals, structural information is of considerable importance. Structural data on $g(r)$ are listed in Table 2.1 (see Fig. 2.19). In view of the importance of the first peak in the $g(r)$ curve, values of $g(r_m)$ for some liquid metals at various temperatures, i.e. the temperature dependence of $g(r_m)$, are given in Figs. 2.20(a), 2.20(b) and 2.20(c).

2.4. THE STRUCTURE OF LIQUID ALLOYS

Theoretical treatments for the structure of a liquid alloy (even a binary alloy) can only be achieved with considerable difficulty. In a binary alloy of components 1 and 2, three pair distribution functions, i.e. the partial distribution functions $g_{11}(r)$, $g_{12}(r)$, and $g_{22}(r)$, are required for a complete description of its structure.

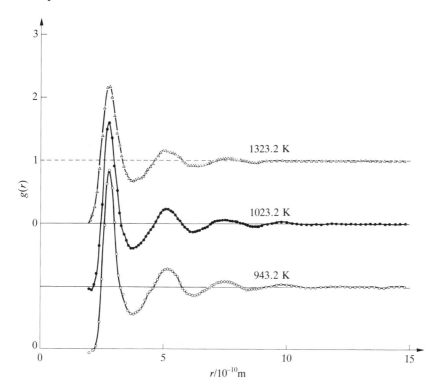

FIG. 2.17. Temperature dependence of the pair distribution function for liquid aluminium.

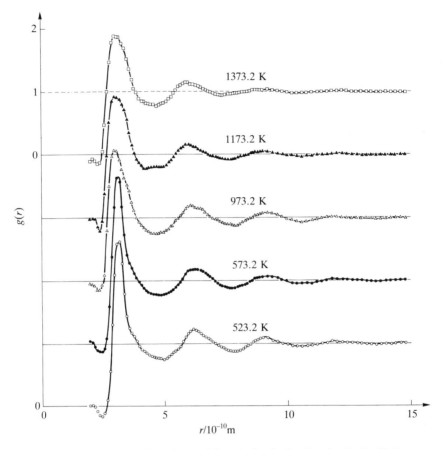

FIG. 2.18. Temperature dependence of the pair distribution function for liquid tin.

In terms of general notation, the partial pair distribution function $g_{\alpha\beta}(r)$ corresponds to the probability of finding an atom β at a distance r from an origin atom α. In other words, if there is an atom α at $r = 0$, the number of atoms β in the spherical shell between distances r and $r + dr$ from the reference atom α is $4\pi r^2 \, dr \, n_0 \, x_\beta g_{\alpha\beta}(r)$, where x_β is the atomic fraction of atoms β. The partial distribution function of a binary system is defined by

$$g_{\alpha\beta}(r) = 1 + \frac{1}{2\pi^2 n_0 r} \int_0^\infty Q[S_{\alpha\beta}(Q) - 1] \sin(Qr) \, dQ \qquad (2.14)$$

where $S_{\alpha\beta}(Q)$ represents the partial structure factor.

T ABLE 2.1. *Structural information on the main peaks of $g(r)$ for liquid metals near their melting points*

Metal	T (K)	r_0 (10^{-10} m)	r_m (10^{-10} m)	$g(r_m)$	a [a]	$\dfrac{r_m}{a}$	$\dfrac{r_0}{r_m}$
Na	378	2.92	3.68	2.42	3.46	1.06	0.79
K	343	3.60	4.56	2.35	4.28	1.07	0.79
Fe	1833	1.98	2.56	2.54	2.37	1.08	0.77
Ni	1773	1.88	2.46	2.36	2.32	1.06	0.76
Co	1823	1.88	2.48	2.37	2.33	1.06	0.76
Cu	1423	2.06	2.50	2.76	2.37	1.05	0.82
Ag	1273	2.34	2.82	2.58	2.68	1.05	0.83
Au	1423	2.34	2.80	2.77	2.66	1.05	0.84
Mg	953	2.52	3.10	2.46	2.95	1.05	0.81
Cd	623	2.54	3.00	2.82	2.86	1.05	0.85
Hg	238	2.62	3.00	2.71	2.90	1.03	0.87
Al	943	2.28	2.78	2.83	2.66	1.05	0.82
In	433	2.70	3.14	2.66	3.00	1.05	0.86
Tl	588	2.74	3.22	2.75	3.11	1.04	0.85
Sn	523	2.68	3.14	2.62	3.05	1.03	0.85
Pb	613	2.76	3.26	3.07	3.18	1.03	0.85
Sb	933	2.58	3.26	2.31	3.15	1.03	0.79
Bi	573	2.78	3.34	2.56	3.26	1.02	0.83
Zn	723	2.16	2.66	2.42	2.52	1.06	0.81
Ga	323	2.38	2.78	2.62	2.67	1.04	0.86

[a] Average interatomic distance = (atomic volume/Avogadro number)$^{1/3}$. Data for $g(r)$ are from Waseda (1980d).

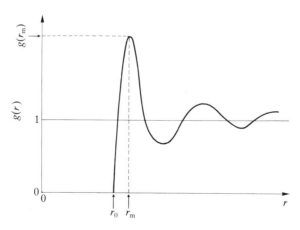

FIG. 2.19. Plot of $g(r)$ curve vs. distance. r_0 and r_m denote the beginning and first peak values of radial distance in the $g(r)$ curve, respectively. $g(r_m)$ denotes the value of $g(r)$ at r_m.

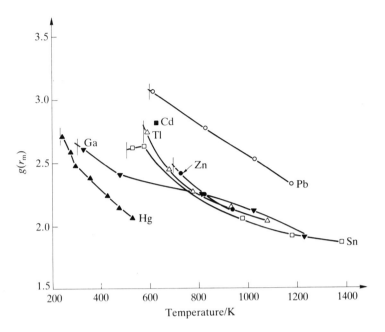

FIG. 2.20(a). Temperature dependence of $g(r_m)$ for several liquid metals.

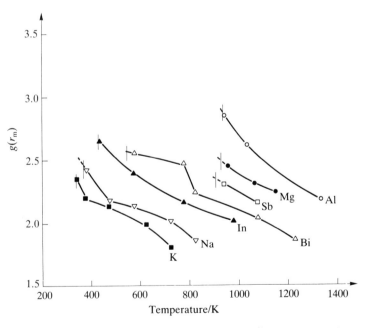

FIG. 2.20(b). Temperature dependence of $g(r_m)$ for several liquid metals.

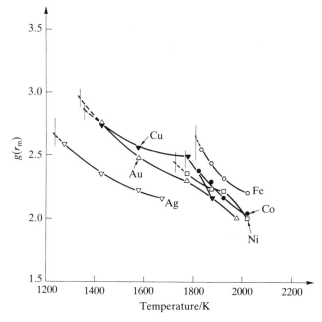

FIG. 2.20(c). Temperature dependence of $g(r_m)$ for several liquid metals.

The structure factor of a binary alloy system obtained from scattering experiments, i.e. the total structure factor $S(Q)$, can be expressed in terms of its three partial structure factors:

$$S(Q) = W_{11}S_{11}(Q) + W_{22}S_{22}(Q) + 2W_{12}S_{12}(Q) \qquad (2.15)$$

where

$$W_{\alpha\beta} = x_\alpha x_\beta f_\alpha f_\beta / \langle x_\alpha f_\alpha + x_\beta f_\beta \rangle^2$$

Here x is the atomic fraction, and f is the atomic scattering factor.

Examples of the structure factors obtained experimentally are shown in Figs. 2.21 and 2.22. The first peak in the S_{Al-Mg} curve lies midway between the first peaks in the S_{Al-Al} and S_{Mg-Mg} curves. By contrast, the first peaks in the S_{Ag-Ag} and S_{Ag-Sb} curves nearly overlap; the peak in the S_{Ag-Sb} curve does not lie halfway between the peaks in the S_{Ag-Ag} and S_{Sb-Sb} curves as would be expected for a random mixture. This suggests that some kind of compound or ordering is present in silver–antimony binary liquid alloys.

The three partial structure factors $S_{\alpha\beta}(Q)$, and the three pair distribution functions $g_{\alpha\beta}(r)$, for liquid iron–phosphorus alloys (metal–metalloid system) obtained by an anomalous scattering technique are shown in Fig. 2.23. On the

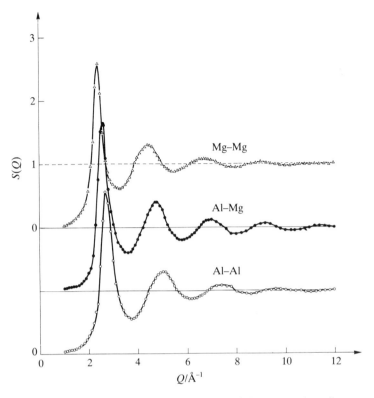

F ig. 2.21. Partial structure factors of liquid aluminium–magnesium alloys.

basis of his experimental results, Waseda (1980e) has discussed the structure of liquid iron–phosphorus alloys and suggested that the fundamental configuration of atoms at the near-neighbour distances in such alloys is similar to that of crystalline Fe_3P. However, he states that the atomic configuration in liquid iron–phosphorus alloys depends mainly on a disordered distribution of iron atoms, similar to that in the dense random-packing model proposed by Bernal (1959) based on tetrahedral units in which the phosphorus atoms occupy vacant spaces left between iron atoms.

The partial structure factors for a number of binary liquid alloys have been computed from observed scattering data. However, it should be kept in mind that these calculations generally contain some assumptions and approximations. Equation (2.15) is a fairly good approximation for a randomly mixed alloy free of any compound-forming characteristics. The total structure factor, or the total interference function, $S(Q)$, for a compound-forming alloy, occasionally has a sub-peak below the first peak. As an example, the experimental data for the total interference functions of liquid copper–indium

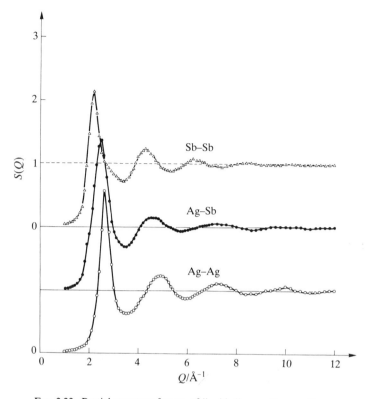

FIG. 2.22. Partial structure factors of liquid silver–antimony alloys.

alloys are shown in Fig. 2.24. The peak positions and heights are summarized in Table 2.2.

According to the experimental results of Isherwood and Orton (1972) for the alloy containing 75.0 at. % indium, the main peak of the total interference function is broad and has two resolvable peaks (a feature which is associated with alloys of Group I–IV elements). In contrast, at the lowest indium concentration of 25.5 at. %, the main peak position is close to that of copper, but there is also a subsidiary maximum close to the peak position of pure indium. Another feature of the pattern from this alloy is the unusually high first main peak. Thus, the total interference function $S(Q)$ indicates a characteristic feature of some liquid alloys. The $S(Q)$ curve for a random mixing alloy would be shaped as those for simple liquid metals.

Finally, as an example of some structural information for liquid binary mixtures, the experimental data for liquid iron–silicon alloys are shown in Figs. 2.25–2.29 (Kita, Zeze, and Morita 1982). According to *Kita et al.*, in the concentration range up to 40 at. % Si, the overall features of $S(Q)$ and $g(r)$

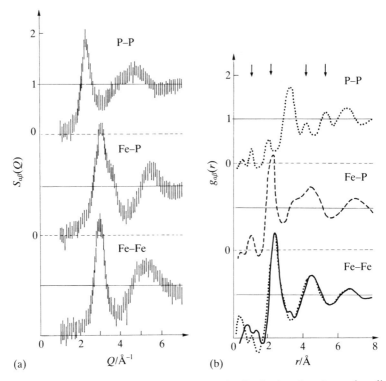

$S_{\alpha\beta}(Q)$

$g_{\alpha\beta}(r)$

P–P

Fe–P

Fe–Fe

P–P

Fe–P

Fe–Fe

(a) $Q/\text{Å}^{-1}$ (b) $r/\text{Å}$

FIG. 2.23. Partial structure factors and partial pair distribution functions of a liquid iron–phosphorus alloy containing 25 % metalloid. The arrows indicate the supposed positions of spurious ripples due to the termination effect (after Waseda 1980e). On the basis of the data of the partial structure factors, the partial pair distribution functions are estimated.

together with Q_1, $S(Q_1)$, W_f, r_1, and Z (see Figs 2.28 and 2.29) are practically equivalent to those for pure liquid iron. For the alloys in the concentration range beyond 50 at. % Si, the first peak of the $S(Q)$ curve becomes broader and more asymmetric, while the first peak of the $g(r)$ curve becomes narrower, and subsequent peaks more strongly damped. Further, the coordination number Z decreases. From these experimental results, they deduced that the structure of liquid iron–silicon alloys remains nearly close-packed up to 40 at. % Si and that it then gradually changes to a more open, lower-coordinated structure at higher silicon levels.

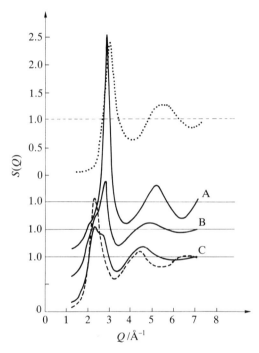

FIG. 2.24. Total structure factors (or total interference functions) (after Isherwood and Orton 1972): dotted line, copper at 1423 K (Breuil and Tourand); A, experimental result for alloy containing 25.5 at. % indium at 973 K; B, experimental result for alloy containing 50.0 at. % indium at 973 K; C, experimental result for alloy containing 75.0 at. % indium at 973 K; broken line, indium at 771 K (Orton and Smith).

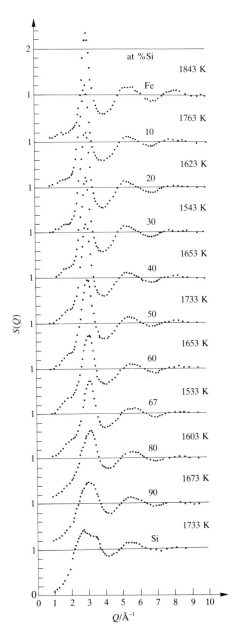

FIG. 2.25. Total structure factors $S(Q)$ for liquid iron–silicon alloys held at super-heat temperatures about 50 K above their liquidus temperatures (after Kita, Zeze, and Morita 1982).

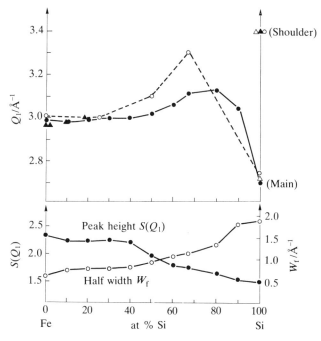

FIG. 2.26. Composition dependence of position Q_1, the height $\check{S}(Q_1)$, and the half width W_f of the first peak of $S(Q)$ for liquid iron–silicon alloys (after Kita *et al.* 1982). ▲ △ Waseda *et al* (▲ 1873 K, △ liquidus + 50 K); ○ Vatolin *et al.* (liquidus + 30 K); ● Kita *et al.* (liquidus + 50 K).

TABLE 2.2. *The peak positions and heights of total structure factors for liquid copper–indium alloys.*

Composition	Total structure factors			
	First peak positions: main and (subsidiary) Q_1 (Å^{-1})	First peak heights: main and (subsidiary)	Second peak position Q_2 (Å^{-1})	Total radial distribution function at first peak position (Å)
Cu	3.00	2.41	5.47	2.60
Cu–25.5 at. % In	(2.13) 2.84	(0.62) 4.07	5.27	2.75
Cu–50.0 at. % In	(2.51) 2.82	(1.33) 1.90	4.90	3.00
Cu–75.0 at. % In	2.32 (2.76)	1.51 (1.38)	4.52	3.20
In	2.28	2.08	4.40	3.30

After Isherwood and Orton (1972).

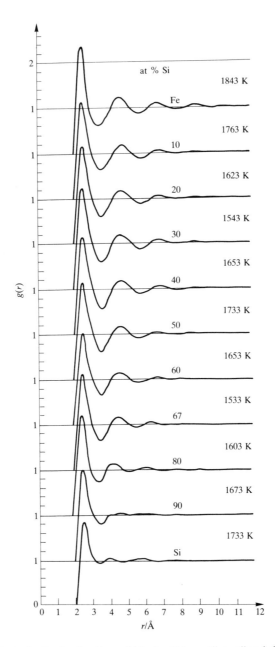

FIG. 2.27. Total pair distribution functions $g(r)$ for liquid iron–silicon alloys held at temperatures about 50 K above their liquidus values (note scale) (after Kita *et al.* 1982).

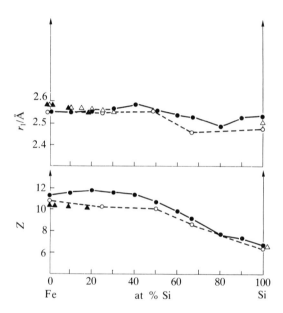

FIG. 2.28. Composition dependence of the nearest-neighbour distance r_1, i.e. the first peak position in r.d.f. curves, and the coordination number Z for iron–silicon alloys (after Kita *et al.* 1982). ▲△ Waseda *et al.* (▲ 1873 K, △ liquidus + 50 K); ○ Vatolin *et al.* (liquidus + 30 K); ● Kita *et al* (liquidus + 50 K).

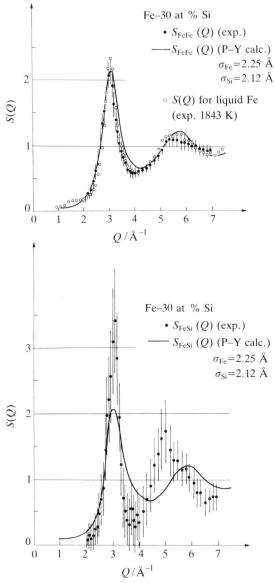

FIG. 2.29. Partial structure factors $S_{FeFe}(Q)$ and $S_{FeSi}(Q)$ for liquid Fe–30 at. % Si alloy, evaluated from $S(Q)$, for Fe–10, 30 and 50 at. % Si alloys by a concentration method together with those calculated from the P–Y (Percus–Yevick) equation for a random mixture of hard spheres, together with $S(Q)$ for liquid iron obtained by experiment. The vertical lines indicate maximum degree of uncertainties which are an inevitable consequence of experimental error (after Kita *et al.* 1982).

Numerical values of $S(Q)$ and $g(r)$ for liquid metals (50 elements) at various temperatures, and of the partial structure factors for liquid binary alloys (25 systems), are provided in Waseda (1980*d*). Figures 2.6–2.18 and 2.20–2.22 are drawn using the experimental data of Waseda (1980*d*).

3

DENSITY

3.1. INTRODUCTION

In discussing the nature and behaviour of liquid metals and alloys, density is an indispensable basic quantity or property. From a practical point of view, density data for liquid metals and alloys provide essential information for topics ranging from mass balance calculations in refining operations or the kinetics of slag/metal reactions to thermal natural convection phenomena in furnaces and ladles. Similarly, the physical separation of liquid metals and overlaying slags, and the terminal velocity of non-metallic inclusions through liquid metals, are essentially dominated by differences in densities between the two phases. As a final example of the importance of density, a detailed knowledge of volume changes in metals and alloys at their melting points is of critical importance in the understanding of solidification processes.

From a more fundamental standpoint, the density or specific volume of a liquid is required in the description of the radial distribution function. Since determinations of this and all of the basic physical properties of liquid metals and alloys (e.g. viscosity, surface tension) require density data, the property of density is of primary importance.

It is not surprising, therefore, that the densities of liquid metals and alloys have long been of interest from both the technological and scientific points of view. Density measurements have been carried out over the past hundred years on a number of liquid metals and binary alloy systems.

However, at present, reliable density data exist only for the common low-melting-point metals and their alloys. More accurate density data on various liquid metals and alloys are still needed. Unfortunately, it is impossible to obtain accurate and reliable experimental data on the density of all liquid metals and alloys, since some of them are either too chemically reactive, too refractory, or too scarce. As a result, density values must be predicted for metals and alloys for which experimental data are lacking. To obtain reliable estimates of density, theoretical studies of density are required. Such an approach to density has been very limited to date. Indeed, densities are generally provided merely as experimental input data for the computation of other physical properties.

3.2. METHODS OF DENSITY MEASUREMENT

Various methods exist for measuring the density of liquid metals and alloys. The essential points of these techniques are described below under the following subheadings:

(a) Archimedean Method (Direct and Indirect),
(b) Pycnometric Method,
(c) Dilatometric Method,
(d) Maximum Bubble Pressure Method,
(e) Manometric Method,
(f) Liquid Drop Method (Sessile Drop Method, Levitation Method),
(g) Gamma Radiation Attenuation Method.

To obtain accurate and reliable density data for liquid metals and alloys, it is necessary to understand the special features of each method of density measurement and to employ the most appropriate one for the system of interest.

3.2.1. Archimedean Method

This technique is based on the well-known principle of Archimedes, and involves its direct, and indirect, application.

3.2.1.1. Direct Archimedean Method

A solid sinker or bob of known weight (in vacuum or, more generally, air) w_1 is suspended by a wire attached to the arm of a balance. When the sinker is immersed in the liquid metal specimen, as shown in Fig. 3.1(a), a new weight w_2, or an apparent loss of weight $\Delta w \, (= w_1 - w_2)$, is observed. The difference in the two weights, i.e. the apparent loss of weight of the immersed sinker, originates mainly in the buoyant force exerted by the liquid metal sample. The

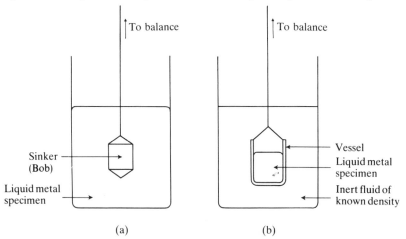

FIG. 3.1. Outline of the Archimedean method.

density of the liquid metal specimen, ρ, is given by

$$\rho = \frac{\Delta w + s}{g(V+v)}$$

$$s = 2\pi r \gamma \cos \theta \qquad (3.1)$$

where g is gravitational acceleration, V is the volume of the sinker, v is the volume of the immersed suspension wire, s is the surface tension correction, i.e. the force acting against the suspension wire of radius r in a liquid for which the surface tension is γ and the contact angle between the wire and the liquid is θ.

From the standpoint of metrology, and in order to obtain very accurate data, the volumes of the sinker and the immersed suspension wire or rod must always be corrected for thermal expansion to operating temperature, since their volumes are, in general, determined experimentally at room temperature. The weight loss of the sinker must also be corrected for surface tension. The effects of s and v in eqn (3.1) can be minimized by using a fine suspension wire, but this is not always practicable in the case of density determinations of metallic liquids at high temperatures. Some workers such as Kirshenbaum and Cahill (1960, 1962a), and Kirshenbaum, Cahill, and Grosse (1961) have evaluated the surface tension correction from the data of γ and θ. According to Veazey and Roe's report (1972), the magnitude of s is of the order of 0.5 per cent of the total buoyancy force Δw for typical metals and suspensions.

Adachi, Ogino, and Kawasaki (1963), Mackenzie (1956; 1959), Martinez and Walls (1973), and Morita, Ogino, Kaitoh, and Adachi (1970) have eliminated the surface tension correction experimentally by using two sinkers of different size. These are suspended by wires of the same diameter. Density measurements of metallic liquids have frequently been made by this latter procedure (the 'double-sinker method') because of a lack of accurate data on surface tensions and contact angles. In addition, the effect of v can be eliminated if the two sinkers are immersed to the same depth within the liquid metal specimen.

From the standpoint of metallurgy, the most important problem with this method is one of materials. Graphite, Pyrex, refractory metals (e.g. molybdenum, tungsten) or metals coated with alumina (Al_2O_3) and zirconia (ZrO_2) can be used for the sinker. While metals are the most suitable for the suspension wire, problems of chemical reaction between the metal wire and the liquid metal sample may occur. Similarly, it is presently very difficult to coat a fine wire with metal oxides such as alumina and zirconia. Consequently, the sinker is frequently suspended by a rod rather than a thin wire. In such a case, the effects of s and v, and the errors arising from deposition of metal vapour on the suspension rod become considerably larger.

In the case of the direct Archimedean Method, a large volume of the liquid metal specimen is required. Nevertheless, numerous workers have employed this technique for density measurements of liquid metals and alloys because it is relatively simple from the experimental point of view and allows for continuous measurements over wide temperature ranges.

Kirshenbaum, Cahill, and Grosse (1962) have reported density values for silver up to 2450 K with an accuracy of $\pm$ 0.1 per cent. Berthou and Tougas (1968) have estimated an error of approximately 0.2–0.3 per cent for thallium and tin up to 1073 K. Hiemstra and co-workers (Hiemstra, Prins, Gabrielse, and Zytveld 1977) have estimated the experimental uncertainty to be about 0.1 per cent for calcium, strontium, and barium up to 1223 K.

3.2.1.2. Indirect Archimedean Method

In this technique, a liquid metal specimen is contained in a vessel, which is weighed while immersed in an inert fluid of established density, as shown in Fig. 3.1(b).

Volume changes during melting can be observed by this method. Gebhardt, Becker, and Dorner (1953) have determined density values for aluminium and aluminium–copper alloys from 373 to 1173 K using this technique. However, the method involved great difficulty, since an accurate knowledge of the other physical properties as well as the density of the inert liquid (oil or molten salt) is also required at the given temperature. Furthermore, it is also extremely difficult to obtain fluid materials which are inert to both the liquid metal specimen and to the vessel or crucible. Nevertheless, an advantage of this technique is that it provides continuous measurements of density from solid to liquid, or from liquid to solid.

3.2.2. Pycnometric Method

A liquid metal specimen is filled in a vessel or pot of known volume, as shown in Fig. 3.2. Upon freezing, the solid metal specimen contained within the pot is weighed at room temperature. One of the characteristics of the pycnometric method is that absolute density determinations are possible.

The principle and construction of the pycnometer are both quite simple. In the case of density measurements of metallic liquids, however, this technique has a problem in respect of materials for the vessel. For precise measurements of density, the volume of the pycnometer must be determined accurately. The following characteristics are required for the materials of a pycnometer: (a) excellent machinability; (b) no reaction with the liquid metal specimen; (c) refractoriness; (d) a low and well-known coefficient of thermal expansion. In practice, no materials exist which have precisely such properties. Consequently, refractory materials such as graphite, quartz, metals (e.g. tantalum) and metal oxides (e.g. alumina, zirconia) have been used for the pycnometer.

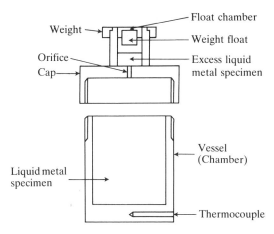

FIG. 3.2. Pycnometer (see e.g. Thresh, Crawley, and White 1968; Crawley 1974).

Crawley has investigated the experimental errors associated with this technique in detail. According to his review (Crawley 1974), these are as follows:

(a) inaccuracies in the measured height and diameter of the pycnometer (± 0.02 per cent),
(b) errors arising from thermal expansion of the pycnometer (± 0.02 per cent),
(c) errors in temperature measurements.

Crawley and White (1968) and Cook (1961) have examined the possibility of incomplete filling of the pycnometer at the sharp corners due to surface tension effects. Their experimental results showed that no detectable errors arose from any such surface tension effects. Density data obtained by this method are considered to be accurate to within ± 0.05 per cent at 800 K.

In summary, pycnometer methods can provide the most accurate density data for low-melting-point metals. Accurate density measurements of metallic liquids at high temperatures (above 1300 K) are considerably more difficult because of limitation in vessel materials. Pycnometric techniques are unsuitable for the continuous measurement of density since a separate filling and weighing is necessary at each temperature.

3.2.3. Dilatometric Method

In this technique for density measurement, an accurately weighed specimen is contained in a dilatometer. This is a calibrated vessel fitted with a long, narrow neck. Changes in volume of the specimen with temperature are continuously monitored through changes in liquid meniscus height. By calibrating the dilatometer beforehand with a liquid of known density, e.g. mercury, the specimen's density and thermal expansivity are obtained. The merits of this method are that large amounts of specimen are not needed and that it provides continuous density measurements. As a result, dilatometric methods have proved popular.

While accurate density measurements require precise measurements of liquid volume, these are sometimes difficult for liquid metals and alloys owing to their meniscus shapes. Consequently, liquid meniscus effects must be corrected. Obviously, the volumetric uncertainty due to the meniscus effect can be decreased by employing a dilatometer with a fine-bore neck. Morita, Iida, and Matsumoto (1985) examined the effect of meniscus on the total volume of liquid specimen, using the Pyrex dilatometer shown in Fig. 3.3. The experimental results demonstrated that the differences in height of the surface of the liquid specimen due to meniscus shape (i.e. Δh_{me} in the capillary tube in Fig. 3.4) were 0.44 mm for distilled water and 0.35 mm for mercury at 290 K, respectively. In this case, the volumetric error arising from the liquid meniscus was, therefore, estimated to be within ± 0.02 per cent. In such experiments, careful degassing of the liquid metal specimen is essential, particularly for

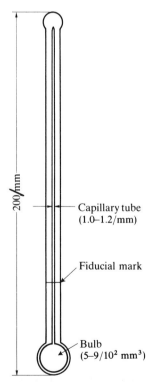

200/mm

Capillary tube
(1.0–1.2/mm)

Fiducial mark

Bulb
(5–9/10^2 mm^3)

FIG. 3.3. Cross section of a dilatometer with a narrow capillary tube (after Morita, Iida, and Matsumoto 1985).

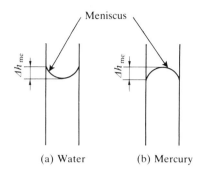

FIG. 3.4. Meniscus shapes in a capillary tube.

those dilatometers with narrow capillary necks. From a metallurgical point of view, the choice of materials for the dilatometer is extremely important. The same material characteristics as those for a pycnometer are needed for the dilatometer.

For transparent dilatometers such as Pyrex or quartz, it is possible to observe the position of the liquid meniscus directly using a cathetometer. On the other hand, in the case of an opaque dilatometer of metal oxide, the meniscus level can be measured indirectly using an electrical contact method (Keskar and Hruska 1972).

When an inert liquid, such as mineral oil, is used as the meniscus-indicating medium in the capillary tube of the dilatometer, measurements of density from solid to liquid (or from liquid to solid), of the specimen's volume change on fusion (Perkins, Geoffrion and Biery 1965), and of volume change on mixing, using a U-shaped cell (Kleppa 1960) can be obtained.

Density data obtained by dilatometric methods are very accurate, and the direct observation method provides particularly accurate and reliable density values. According to Williams and Miller (1950) and McGonigal and Groose (1963), experimental errors in dilatometric techniques were $\pm$ 0.2 per cent for indium up to 573 K and $\pm$ 0.4 per cent for arsenic up to 1323 K, respectively. Knight (1961) and Perkins *et al.* (1965) have determined density data for highly reactive and toxic metals such as plutonium and cerium, using a form of dilatometer.

3.2.4. Maximum bubble pressure method

When an inert gas is passed through a capillary tube immersed to a certain depth in a liquid, bubbles of gas will detach from the tip of the capillary (Fig. 3.5). The density of the liquid specimen can be determined by measuring the over-pressure required to just detach a bubble of the inert gas from the tip of the capillary. The maximum bubble pressure P_m at depth h is equal to the sum of the pressure needed to maintain the column of liquid, $\rho g h$, and the pressure

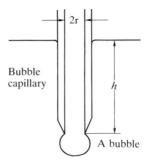

FIG. 3.5. A bubble tube with a conically ground tip.

needed to create a bubble's new surface, $2\gamma/r$, (see Subsection 5.2.1) so that,

$$P_m = \rho g h + \frac{2\gamma}{r} \qquad (3.2)$$

where γ is the surface tension and r the radius of the capillary tube. This requires that spherical-shape bubbles of radius r detach from the tips of the capillary. In other words, corrections must be made for non-spherical bubbles. The second term in the above equation, the surface tension effect, can be eliminated by making determinations of the maximum bubble pressures at different depths of immersion. Hence, if the maximum bubble pressures are P_{m1} and P_{m2}, at the depths of immersion h_1 and h_2, respectively, the density of the liquid specimen is given by

$$\rho = \frac{P_{m1} - P_{m2}}{g(h_1 - h_2)} \qquad (3.3)$$

It is obvious from this equation that the maximum bubble pressure P_m and depth of immersion h or displacement $\Delta h\,(= h_1 - h_2)$ require accurate measurements for reliable density data. Furthermore, a correction must be made for any expansion of the capillary tube.

   This technique allows density measurement to be carried out at high temperatures and over wide temperature ranges. Many workers have employed the method for density measurements of metallic liquids. However, it is not as accurate as the pycnometric method, since precise measurements of maximum bubble pressure are technically difficult. Similarly, the authors have shown that non-wetting effects in gas/liquid metal systems can lead to spreading away from an inside nozzle perimeter and across the surfaces of the capillary tube (Irons and Guthrie 1981). These can reduce P_m values. In addition, a rather large volume of liquid sample is needed.

Lucas (1970) has determined density values for various liquid metals from their melting points up to 2100 K using this method. On the basis of his detailed examination, he concluded that the accuracy of the technique was approximately $\pm$ 0.3 per cent. A typical maximum bubble pressure experimental set-up is given by Veazey and Roe (1972).

For further details of this method, see Subsection 5.2.2.2.

3.2.5. Manometric Method

This method requires an application of the U-tube manometer. The manometric method consists of measuring the difference in the heights of two columns of the liquid specimen contained within a U-tube. The height differential is maintained through a gas pressure difference between the two arms of the U-tube. This pressure difference is measured using another manometer, or by electric contact. The manometric technique has characteristics that are common to the maximum bubble pressure method in measuring the pressures, and common to the dilatometric method in the measurement of liquid levels. The sources of error in the manometric method are similar to those in the dilatometric and the maximum bubble pressure method.

Recently, densities of liquid iron and iron–carbon alloys have been measured using the manometric method based on an x-ray transmission technique (Ogino, Nishiwaki, and Hosotani 1984a, 1984b).

3.2.6. Liquid Drop Method

The liquid drop methods involve photographing profiles of a liquid metal drop. Following solidification, the metal specimen is then weighed and the volume of the liquid metal drop calculated through geometrical analysis of the photographs.

Two methods are used for maintaining a stable liquid metal drop: the first employs the sessile drop method, the second uses levitation techniques. These methods can be used for high-temperature systems up to about 2300 K.

3.2.6.1. Sessile Drop Method

A liquid metal drop resting on a plate or substrate of a smooth horizontal surface acquires a shape such as that shown in Fig. 3.6. The volume of this liquid drop can then be calculated from the numerical table of Bashforth and Adams (1883) by determining every size, i.e. $X, \ldots, \theta$, indicated in Fig. 3.6 (see Subsection 5.2.2.1). To obtain accurate values for drop volumes, great efforts have been expended in producing truly symmetrical liquid drops. However, precise volume measurements seem to be difficult. Examples of reported errors in the literature are: $\pm$ 0.25 per cent for zinc up to 823 K by White (1966), and

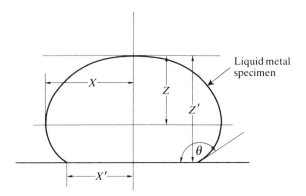

FIG. 3.6. A liquid metal drop on a plate (the sessile drop method).

± 2 per cent for iron, cobalt, and nickel up to 2023 K by Vertman, Samarin, and Filippov (1964).

To prevent a sessile drop of asymmetrical shape forming, Vertman and Filippov (1964; see Crawley 1974) have proposed the so-called 'large drop method'. The large drop method is outlined in Fig. 3.7. A sessile drop surmounting a cylindrically ground and edged vessel is apt to be perfectly symmetrical. As can be seen from Fig. 3.7, this method may be considered to be a combination of the pycnometric and sessile drop techniques. According to Vertman and Filippov, the accuracy of masurements is of the order of 0.5 per cent.

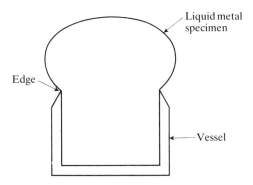

FIG. 3.7. Outline of the large drop method.

3.2.6.2. Levitation Method

The principle of this technique is to float or levitate a metal drop through electromagnetic induction. El-Mehairy and Ward (1963) developed this method in order to eliminate the problem of any chemical reactions between liquid metal drops and ceramic substrates or vessels. Later, Saito and Sakuma (1967), Saito, Shiraishi, and Sakuma (1969), and Adachi, Morita, Kitaura, and Demukai (1970) examined this technique for density measurements of liquid iron and its alloys. They report that the error in the levitation technique arises mainly from length measurements of the sectioned images of a levitated drop. The accuracy attainable with this method is generally inferior to the pycnometric, the dilatometric, and the Archimedean methods. However, levitation techniques do have the advantage of being compatible with highly reactive metallic elements.

Finally, it is worth noting that the liquid drop and the maximum bubble pressure methods not only allow density measurements but also surface tension evaluations.

3.2.7. Gamma Radiation Attenuation Method

The principle of this method is based on the attenuation of a gamma ray (γ-ray) beam passing through matter. The experimental arrangement for the gamma radiation attenuation technique is shown schematically in Fig. 3.8. The γ-ray beam from the source of radiation, e.g. ^{60}Co (Vertman and Filippov 1964, see Crawley 1974), ^{137}Cs (Döge 1966), in a lead container is collimated before entering a liquid metal specimen. After the incident beam is attenuated according to the mass of the liquid metal sample, the intensity of the emergent beam is recorded by a radiation counter. For a γ-ray beam of intensity I_0 penetrating a specimen of length x the emergent radiation intensity I is given

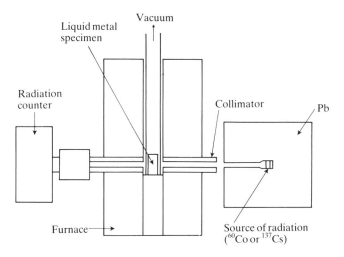

FIG. 3.8. Schematic diagram of the experimental set-up for the gamma radiation attenuation method (see e.g. Döge 1966).

by

$$I = I_0 \exp\left(-\alpha \rho x\right) \tag{3.4}$$

where α is the absorption coefficient per unit mass. In determining the liquid metal's density, values for α can first be obtained from absorption measurements on the metal in its solid form, where its density will be well known or easily measured.

Basin and Solov'ev (1967) have obtained density data for liquid lead using this technique. According to their experimental results, the differences in the density values for liquid lead determined by gamma radiation attenuation and the pycnometric methods were 0.16, 0.10, and 0.05 per cent at the melting points (600.6), 673, and 773 K, respectively (Crawley 1974). This technique, therefore, seems to provide accurate density data. It can be seen from eqn (3.4) that the accuracy of this technique depends on the sensitivity of the radiation counter.

However, one should note that it is still technically difficult to obtain reliable density data at high temperatures using this method. On the other hand, an advantage of the technique is that surface tension effects and chemical contamination of the liquid metal's surface are not involved, since the γ-ray beam penetrates the bulk of the specimen.

3.3. ESTIMATION OF DENSITY

3.3.1. Empirical or semi-empirical equations for the density of liquid metals

Of the several empirical expressions for thermal density variations of liquids, the most familar is the law of Cailletet and Mathias, or the law of rectilinear diameters. This law states that the average density of a liquid and its vapour, or half the sum of the densities of a liquid and its saturated vapour, decrease linearly as the temperature rises up to the critical point. This is equivalent to saying that

$$\frac{\rho_1 + \rho_g}{2} = a - bT \tag{3.5}$$

holds over the entire temperature range of a liquid, where ρ_1 is the liquid's density and ρ_g is the vapour's density, T is the absolute temperature, and a and b are constants. At the critical point, the densities of liquid and saturated vapour are equal, i.e. $\rho_1 = \rho_g = \rho_c$. (Where ρ_c represents the substance's density at its critical temperature T_c). Equation (3.5) can be rewritten in dimensionless form using the critical density ρ_c:

$$D^* = \frac{\rho_1 + \rho_g}{2\rho_c} = 1 + \mu(1 - T^*) \tag{3.6}$$

where D^* is the reduced (or dimensionless) rectilinear diameter, T^* ($\equiv T/T_c$) is the reduced temperature and μ ($\equiv bT_c/\rho_c$) is a constant whose value is considered smaller than unity.

The validity of eqns (3.5) or (3.6) has been confirmed experimentally for a large variety of thermally stable organic and inorganic liquids such as benzene, water, and argon. In the case of liquid metals, few data on critical temperatures and critical densities have been obtained because property measurements at high temperatures and high pressures are difficult and require special materials and techniques. To our knowledge, only mercury and the alkali metals have been heated to, or near, their critical points. Thus, using the law of rectilinear diameters, estimates for critical temperatures and corresponding critical densities of various metals have had to be made as an alternative solution.

Grosse (1961a) proposed a method for estimating critical temperatures of liquid metals, based on the assumption that the molar entropies of vaporization of all liquid metals, $\Delta_1^g S$ ($= \Delta_1^g H/T$, where $\Delta_1^g H$ is the heat of vaporization per mole), can be expressed as a universal function of the reduced temperature T^*, i.e. that the curve of $\Delta_1^g S$ versus T^* is the same for all metals as that for mercury†.

Later, Dillon, Nelson, and Swanson (1966) obtained experimental data for the critical constants of the liquid alkali metals. Their results demonstrate that a universal relationship can only be regarded as roughly approximate, as seen from Fig. 3.9. However, one can use the correlation to estimate approximate data for the critical temperatures of liquid metals. Consequently, using experimentally determined entropies of vaporization for other metals (e.g. at their normal boiling points), one can read off their approximate critical temperatures from the curves given in Fig. 3.9. Once the critical temperature of a metal is known, its corresponding critical density can be estimated using the law of rectilinear diameters, i.e. by extrapolating the experimental data on densities up to the critical temperature. An example of the density and rectilinear diameter plot is shown in Fig. 3.10. A generalized correlation for plotting reduced density ρ^* ($\equiv \rho/\rho_c$) versus reduced temperature is given in Fig. 3.11.

Since the density of a liquid, ρ_l, is much greater than the density of a vapour, ρ_g, i.e. $\rho_l \gg \rho_g$, over a wide temperature range, one can, using eqn (3.6) write the following expression:

$$\rho^* \approx 2 + \{\rho_m^*/(1 - T_m^*) - 2\}(1 - T^*) \tag{3.7}$$

where ρ_m^* represents the reduced density (ρ_m/ρ_c) at the metal's melting point T_m. McGonigal (1962) has presented an alternative method for estimating

---

† At that time only the critical constants for mercury had been determined experimentally.

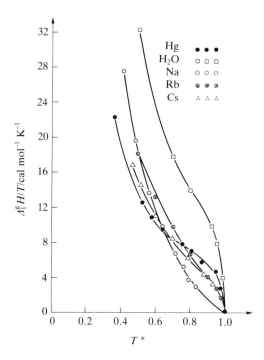

FIG. 3.9. Variation of entropy of vaporization with reduced temperature for alkali metals, mercury, and water (after Dillon, Nelson, and Swanson 1966).

critical densities using the dimensionless variable Δ. On the basis of eqn (3.5) or (3.6), this dimensionless quantity was defined as

$$\Delta \equiv \left(\frac{dD^*}{dT^*}\right) \bigg/ \left(\frac{D^*}{T^*}\right) \tag{3.8}$$

Correlations between $\Delta^*(\equiv \Delta/\Delta_c$, where Δ_c is the value of Δ at the critical point) and T^* are shown in Fig. 3.12. McGonigal reports that the agreement among the liquid metals shown in Fig. 3.12 can be considered to be good, excepting the Δ^* values of the Group IIB metals. Δ^* can be expressed mathematically as

$$\Delta^* = \frac{T^*}{1 + \mu(1 - T^*)} \tag{3.9}$$

However, since μ is not a universal constant for all liquid metals, Δ^* cannot be regarded as a universal quantity.

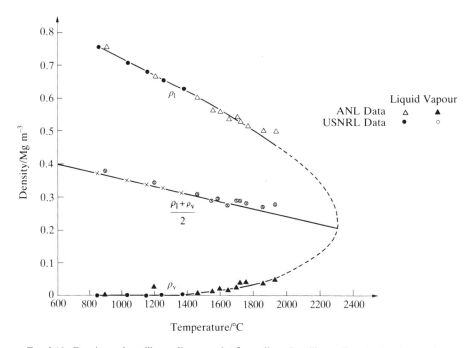

FIG. 3.10. Density and rectilinear diameter plot for sodium. Rectilinear diameter line is given by $(\rho_1 + \rho_v)/2$, where the experimental points are given as follows: ⊗ ⊗ ⊗, Argonne National Laboratory data; x x x, US Naval Research Laboratory data. Estimated critical constants: $\rho_c = 0.21 \times 10^3\ \mathrm{kg\,m^{-3}}$, $T_c = 2573\ \mathrm{K}$, $P_c = 3.55 \times 10^7\ \mathrm{Pa}$ (after Dillon *et al.* 1966).

It has been known since the last century that the following relationship between the coefficient of volume thermal expansion at 293 K, α_0, and the melting temperature T_m holds for solid metals:

$$\alpha_0\, T_m \approx 0.06 \qquad\qquad (3.10)^\dagger$$

Using this equation, Steinberg (1974) derived an expression for Λ, showing that

$$\frac{\Lambda\, T_b}{D_{00}} \approx -0.23 \qquad\qquad (3.11)$$

---

† The following relationships hold for liquid metals (Iida and Morita 1978):

$$\alpha\, U_c \approx A,\ \text{or}\ \Lambda U_c \approx -A\rho$$

where α is the thermal expansivity, U_c is the cohesive energy, and A is a constant.

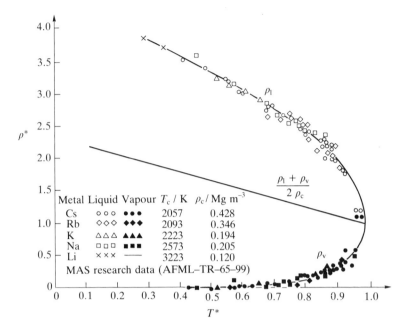

Metal	Liquid	Vapour	T_c / K	ρ_c/ Mg m^{-3}
Cs	∘∘∘	•••	2057	0.428
Rb	◇◇◇	◆◆◆	2093	0.346
K	△△△	▲▲▲	2223	0.194
Na	□□□	▪▪▪	2573	0.205
Li	×××	——	3223	0.120

MAS research data (AFML–TR–65–99)

FIG. 3.11. Generalized correlation of reduced density with reduced temperature (after Dillon *et al.* 1966).

where T_b is the boiling point and D_{00} ($\equiv \rho_m - \Lambda T_m$) is the density determined by extrapolating ρ_m to absolute zero. The data plotted in Fig. 3.13 indicate that this correlation is approximately true for a large number of metals, excepting those in Groups VIII and IIB. Equation (3.11) has been used to estimate Λ for high-melting-point metals for which no experimental data are available (see Table 3.3).

Iida and Morita (1978) have also derived a simple expression for the temperature dependence of liquid metal density Λ. The analysis gives (see the end of this Chapter for derivation):

$$\Lambda \approx -2.8 \times 10^{-5} \xi M^{\frac{1}{2}} \qquad (3.12)$$

Here M is the metal's atomic weight and ξ is a parameter related to the repulsive energy of the pair interaction potential.† The data plotted in Fig. 3.14 show that this relationship is approximately true for various liquid metals, although iron, nickel, cobalt, and copper deviate somewhat from the relationship. For example, the measured value of Λ for mercury, and that calculated through the use of eqn (3.12), are 24×10^{-4} Mg m^{-3} K^{-1} and

† There exists a simple relationship between the parameter ξ and the repulsive exponent n of the pair interaction (Matsuda and Hiwatari 1973): $\xi \approx 0.15n$ (see Subsection 4.44).

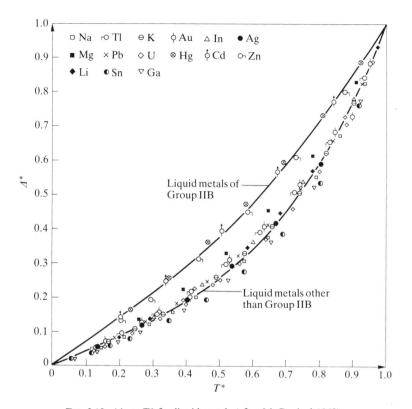

FIG. 3.12. Δ^* vs. T^* for liquid metals (after McGonigal 1962).

$26 \times 10^{-4}\, \text{Mg}\ \text{m}^{-3}\, \text{K}^{-1}$, respectively, these lying outside the range of Λ covered in Fig. 3.14.

Most of the physical and chemical properties of liquid metals (e.g. atomic volumes, compressibilities, ionization potentials) vary periodically with atomic number. In Fig. 3.15, the temperature dependence of the density of liquid metals is shown as a function of their atomic number. It can be seen that there is a periodic variation in Λ values. Broadly speaking, close-packed metals in the solid state lie on the peaks of the curve, while b.c.c. metals lie in the valleys. Evidently, a correlation exists between the value of Λ in the liquid state and the crystal structures in the solid state at temperatures close to their melting points. As is clear from Fig. 3.15, values of Λ for the b.c.c. metals, zirconium, niobium, molybdenum, hafnium, tantalum, tungsten, and thorium, estimated by Steinberg (1974) fit into this periodic relationship very well.

Finally, another relationship, known as Thiesen's equation, has been suggested:

$$\rho_1 - \rho_g \propto (T_c - T)^{\frac{1}{3}} \qquad (3.13)$$

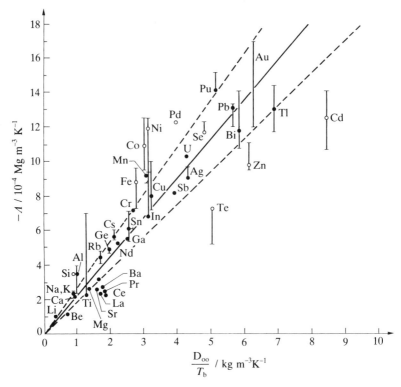

FIG. 3.13. Correlation of Λ with D_{00}/T_b for 41 elements. Open circles represent semi-metals and Group VIII and IIB metals. Mercury and platinum points fall outside the range of the figure. Their ordinates are 24.2 and 28.8 and their abscissas 22.6 and 6.05, respectively. The broken lines represent ± 20 per cent error band (after Steinberg 1974).

3.3.2. Theoretical analyses of density

As mentioned above, there has been little theoretical interest in the property of 'density' itself. Density has only been discussed as a part of studies of the structure and physical properties of liquids. However, density, or packing density (packing fraction) η, is a parameter of fundamental importance in explaining the behaviour and properties of liquids. These properties, viscosity, diffusivity, and liquid–solid phase transition, can be interpreted on the basis of the hard-sphere model of liquids.

Wood and Jacobson (1957) were the first to begin computational experiments on packing density. Alder and Wainwright (1957, 1962; Wainwright and Alder 1958) went on to examine the behaviour of hard-sphere systems as a function of the packing fraction by means of a mathematical model of atomic

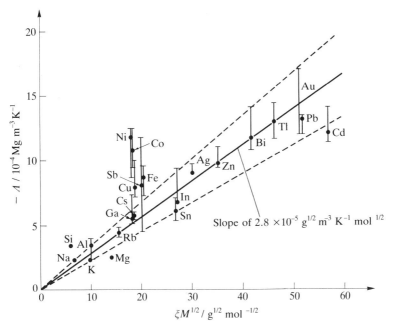

FIG. 3.14. The temperature dependence of density of various liquid metals as a function of $(\xi M^{\frac{1}{2}})$. Mercury falls outside the range of the figure. Its ordinate is 24×10^{-4} $Mg\,m^{-3}\,K^{-1}$ and its abscissa 93. The broken lines represent ± 20 per cent error band (after Iida and Morita 1978; Iida, Fukase, and Morita 1981).

collisions. They studied a two-dimensional system containing 870 hard-disk particles. The simultaneous motions of the particles were obtained through mathematical solutions with the aid of a high-speed digital computer. In their work, the equation of state for the hard-disk particles was set up as a function of the reduced area A^*. This reduced area was defined as being the ratio of the total area of the particle system A to its area at close packing, A_0, i.e. $A^* \equiv A/A_0$. The authors concluded that a fluid† and a solid phase (in a van der Waals-like loop) coexisted in the region of A^* from 1.26 to 1.33 (Alder and Wainwright 1962). Furthermore, they were able to predict accurately the transport coefficients of the hard-sphere fluid as a function of the atoms' packing fractions up to 0.49. According to their computations, transport coefficients changed very rapidly at the highest densities. With increasing η, the

† One-component hard-sphere systems have two different equilibrium phases: a fluid phase and a solid phase (a crystalline phase). Thus there is no distinction between a liquid and a gas in the hard-sphere systems. To separate the liquid state from the gas, attractive forces would be needed in the model (Barker and Henderson 1981).

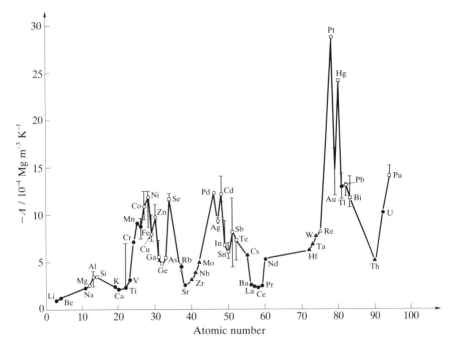

FIG. 3.15. The periodic variation in the temperature dependence of density of various liquid metallic elements (after Iida and Morita 1978). The signs refer to the crystal structures in solid state at temperatures close to melting points: ○ △, close-packed structure; ● ▲ body-centered cubic structure; □, other complex structures. For △ ▲, Λ values are estimated by Steinberg (1974).

fluid viscosity increased rapidly, while self-diffusivity decreased rapidly. These results suggest that the hard-sphere model exhibits the phenomenon of a liquid (fluid)–solid transition, which is equivalent to that of a real substance. This characteristic of hard-sphere systems (the so-called Alder transition) implies that the repulsive force between atoms represents the principal factor governing liquid–solid transformations and atomic arrangements.

The occurrence of a phase transition in the hard-sphere system depends only on packing fraction. However, this phase transition in real liquids depends not only on the packing fraction but also on absolute temperature, because the repulsive interactions between atoms are not perfectly elastic in practice.

Subsequent theoretical considerations by Hoover and Ree (1967, 1968) have indicated that while the value of the packing fraction for hard spheres is 0.49 at the point of freezing, this value is increased to 0.54 at the point of melting (a hysteresis phenomenon). The fluid was found to be stable for all packing fractions equal to, or less than, 0.49.

Over the past twenty years, considerable interest in the structures and properties of amorphous solids or metallic glasses has developed. Woodcock (1976), and Hudson and Andersen (1978) have demonstrated that the hard-sphere fluid can undergo a glass-like transition when it is compressed to a high density. The estimated value of the packing fraction at the glass transition is 0.53. This value is substantially lower than 0.64, the packing fraction of a dense, randomly packed array of hard spheres.

As mentioned, repulsive forces between atoms (ions) in liquids are, in fact, softer than those in the hard-sphere model. Notwithstanding this, several properties of liquid metals can be explained successfully by treating ions as hard spheres. Examples include volume changes on melting, and self-diffusivities, together with the structure factors and pair distribution functions mentioned in Chapter 2.

There has been recent progress in a pseudopotential theory of liquid metals (e.g. Harrison 1966). This has led a better understanding of some physical properties, e.g. electrical resistivity, of the more simple liquid metals, such as sodium or potassium.

At the present time, however, it is impossible to predict reliable density values for liquid metals and alloys because sufficiently accurate forms of the pair potentials between specific ions are not yet available.

3.4. ATOMIC VOLUME OF LIQUID METALS AND ALLOYS

In the theoretical treatment of liquid metals, atomic volumes or molar volumes are frequently used for describing their properties. The atomic volume of a pure metal, V, is defined as the ratio of its atomic weight M to density ρ:

$$V = \frac{M}{\rho} \tag{3.14}$$

For a binary system, the atomic volume V_A is defined by

$$V_A = \frac{x_1 M_1 + x_2 M_2}{\rho_A} \tag{3.15}$$

where x_1, x_2, M_1 and M_2 are the atomic fractions and the atomic weights of components 1 and 2, respectively, and ρ_A is the alloy density. In studies of real alloy systems, the excess volume V^E is frequently used. For a binary alloy system,

$$V^E = V_A - V_{ideal}$$
$$= \frac{x_1 M_1 + x_2 M_2}{\rho_A} - \left(\frac{x_1 M_1}{\rho_1} + \frac{x_2 M_2}{\rho_2} \right) \tag{3.16}$$

In addition, as large differences in the atomic volumes of two metals forming a liquid alloy can have an important effect on its properties, related size factor parameters are also used in theoretical descriptions.

Various qualitative and quantitative investigations have been made to explain the excess volume of a liquid mixture. From the standpoint of thermodynamics, Scatchard (1937) has presented qualitative considerations of V^E, the volume change on mixing at constant pressure. On the basis of Scatchard's treatment, Kleppa (1960), and Kleppa, Kaplan and Thalmayer (1961) have proposed an approximate expression for a volume contribution to the entropy of mixing or a contribution to the excess entropy S^E arising from this excess volume V^E.

$$S^E \approx \left(\frac{\alpha}{\kappa_T} \right) V^E \tag{3.17}$$

$$H^E \approx T \left(\frac{\alpha}{\kappa_T} \right) V^E \tag{3.18}$$

Here α is the thermal coefficient of expansion, κ_T is the isothermal compressibility and H^E is the enthalpy of mixing or the heat of mixing. In the above equations, the value of the excess entropy is considered to arise solely, and excess enthalpy only slightly, from volume changes on mixing.

Predel and Emam (1969) have indicated that there is no correlation between the maximum values of enthalpy of mixing and excess volumes. This is demonstrated in Fig. 3.16. In an investigation of about 40 alloy systems, Crawley (1974) has also noted a lack of correlation between the signs of V^E and thermodynamic properties such as S^E and H^E. On the other hand, Marcus (1977) has demonstrated that there is a good correlation between the signs of V^E and S^E in some equimolar liquid binary alloys (31 alloys) forming solid solutions and eutectics. Kubaschewski and Alcock (1979c) have also examined possible relationships between excess volumes and thermodynamic properties. Their results for 18 binary alloy systems show that a rough connection does exist between entropy and volume changes when liquid metals are mixed. There also appears to be an approximate proportionality between changes in volume and heats of mixing.

Simplistically speaking, a negative V^E value indicates attractive interactions between dissimilar kinds of atoms. In other words, alloy systems showing negative volume changes on mixing have larger mutual-, as opposed to self-, interaction energies. Consequently, excess thermodynamic functions, i.e. S^E and H^E, would be expected to show negative values. Negative V^E values have been observed for compound-forming alloy systems.[†] Some of these systems show large negative V^E values. By contrast, positive values of V^E indicate repulsive interactions between different kinds of atoms, and S^E and H^E are

[†] Using the theories of liquids, the structures and properties of compound-forming alloy systems have been studied (e.g. Tamaki, Ishiguro, and Takeda 1982).

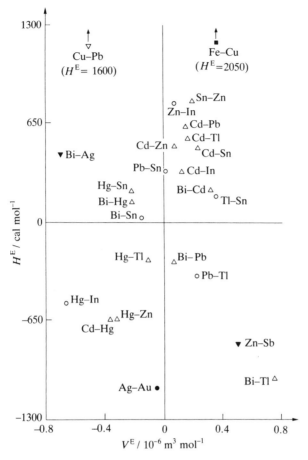

FIG. 3.16. Relation between the maximum values of enthalpy of mixing H^E and excess volume V^E for binary liquid alloys (after Predel and Emam 1969); $\triangle$ Kleppa *et al.*; $\bullet$ Gebhardt *et al.*; $\blacktriangledown$ Sauerwald; $\blacksquare$ Frohberg; $\triangledown$ Malmberg; $\bigcirc$ Predel and Emam. Data for H^E from Kleppa, Hawkins, Hultgren *et al.*, Schürmann and Kaune.

then positive. Alloy systems exhibiting immiscibility (i.e. a miscibility-gap) are reported to show positive V^E and H^E values.

Finally, solid solutions and eutectic-forming alloys have comparable self- and mutual-interaction energies.

At the present time, an adequate amount of excess volume data for various types of liquid alloys is still lacking. In particular, reliable and systematic data for the density of liquid binary alloys are scarce. It is, therefore, difficult to make a detailed comparison of experimental data and theory.

3.5. EXPERIMENTAL DATA FOR LIQUID METALS AND ALLOYS

3.5.1. Experimental densities of pure liquid metals

Table 3.1 lists density data available for pure liquid metals. The data sources for densities of pure liquid metals at their melting points, ρ_m, and their temperature dependence Λ are from Steinberg (1974)[†], In choosing preferred values of ρ_m and Λ from an often large and conflicting body of data, the following points were considered (Steinberg 1974): (a) liquid density measurements made over a wide temperature range were preferred because the wider the range, the more accurate should be the determinations of ρ_m and Λ; (b) wherever possible, the 'Range' quoted in Table 3.1 by Steinberg contains the reasonable extremes of modern experimental values for Λ. The complete range of all data is usually much larger. Table 3.1 also contains the liquid atomic volumes at the melting point V_m, and thermal coefficients of expansion α.

Seydel and Kitzel (1979) have obtained thermal volume expansion data for liquid titanium, vanadium, molybdenum, palladium, and tungsten in the temperature range between melting and boiling using a fast resistive pulse heating method. Their results are given in Table 3.2. Estimated values for high melting-point metals are listed in Table 3.3.

Figure 3.17 shows the periodic variation in atomic volumes. Figure 3.18 shows a relationship between atomic volumes and enthalpies of evaporation or cohesive energies for 43 metallic elements. As can be seen from this figure, with the exception of Group II metals, and the lanthanide and actinide elements, most metals lie within a ± 30 per cent error band of the expression $\Delta_l^g H_m = 3.7 \, V_m^{-1}$. The cohesive energy ($\Delta_l^g H_m$), which is a basic physical quantity in discussing the structures and properties of condensed matter, can be approximately expressed in terms of atomic volumes alone.

It is found experimentally that the temperature dependence of density for liquid metals and alloys is linear. Density data can be adequately represented by the equations

$$\rho = \rho_m + \Lambda(T - T_m) \tag{3.19}$$

$$V = V_m\{1 + \alpha(T - T_m)\} \tag{3.20}$$

However, for aluminium, gallium, and antimony, there is sufficient data to indicate that Λ is not constant over the whole liquid range[‡] In this case, Table 3.1 provides an average value for Λ: $\Lambda\{ \equiv -(\rho_m - \rho_b)/(T_b - T_m)\}$, where ρ_b is the density at the boiling temperature T_b (Steinberg 1974). In general, the observed linear temperature variation of liquid metal densities may not be precisely true. If these measurements had been made over a wider temperature

---

[†] In a review paper by Crawley (1974), the experimental data available on the densities of pure liquid metals are also listed and a detailed appraisal of the data is made.
[‡] Anomalous discontinuities in the temperature dependence of the density of liquid iron have been reported by Morita, Ogino, Kaitoh, and Adachi (1970) and Vertman et al. (1964).

TABLE 3.1. *Information on densities of pure liquid metals*

Metal	ρ_m (10^3 kg m^{-3})	$-\Lambda(\equiv -(\partial\rho/\partial T)_p)$ (10^{-1} kg m^{-3} K^{-1})	Range[a]	V_m (10^{-6} m^3 g-atom^{-1})	$\alpha_m \equiv -\Lambda/\rho_m$ (10^{-4} K^{-1})
Ag	9.33	9.1	9.1–9.7	11.6	0.98
Al	2.38	3.5	2.4–4.0	11.3	1.5
As	5.22	5.4	–	14.4	1.0
Au	17.4	12/17	–	11.3	0.69/0.98
Ba	3.32	2.7	–	41.4	0.81
Be	1.69	1.2	–	5.33	0.71
Bi	10.05	11.8	10.8–14.1	20.80	1.17
Ca	1.36	2.2	–	29.5	1.6
Cd	8.01	12.2	11.4–14.1	14.0	1.5
Ce	6.69	2.3	–	20.9	0.34
Co	7.75	10.9	9.5–12.5	7.60	1.4
Cr	6.29	7.2	–	8.27	1.1
Cs	1.84	5.7	5.5–6.0	72.2	3.1
Cu	8.00	8.0	7.2–10.0	7.94	1.0
Fe	7.03	8.8	7.3–9.6	7.94	1.3
Ga	6.10	5.6	7.3–5.1	11.4	0.92
Ge	5.49	4.9	4.7–4.9	13.2	0.89
Hg	13.69	24.2	–	14.65	1.77
In	7.03	6.8	6.8–9.4	16.3	0.97
K	0.83	2.4	2.2–2.5	47.1	2.9
La	5.96	2.4	–	23.3	0.40
Li	0.518	1.0	–	13.4	1.9
Mg	1.59	2.6	–	15.3	1.6
Mn	5.76	9.2	–	9.54	1.6
Na	0.927	2.35	2.25–2.45	24.8	2.54
Nd	6.69	5.3	–	21.6	0.79
Ni	7.90	11.9	8.7–12.5	7.43	1.51
Pb	10.67	13.2	12.0–13.3	19.42	1.24
Pd	10.49	12.3	–	10.14	1.17
Pr	6.61	2.5	–	21.3	0.38
Pt	18.91	28.8	–	10.31	1.52
Pu	16.65	14.1	14.1–15.2	14.65	0.847
Rb	1.48	4.5	4.1–4.8	57.7	3.0
Sb	6.48	8.2	4.5–11.8	18.8	1.3
Se	4.00	11.7	11.7–12.3	19.7	2.93
Si	2.53	3.5	–	11.1	1.4
Sn	6.98	6.1	5.4–7.1	17.0	0.87
Sr	2.37	2.6	–	37.0	1.1
Te	5.80	7.3	5.2–7.3	22.0	1.3
Ti	4.13	2.3	2.3–7.0	11.6	0.56
Tl	11.35	13.0	11.7–14.4	18.00	1.15
U	17.27	10.3	–	13.78	0.596
V	5.36	3.2	–	9.50	0.60

TABLE 3.1 (continued)

Metal	ρ_m (10³ kg m⁻³)	$-\Lambda(\equiv -(\partial\rho/\partial T)_p)$ (10⁻¹ kg m⁻³ K⁻¹)	Range[a]	V_m (10⁻⁶ m³ g-atom⁻¹)	$\alpha_m \equiv -\Lambda/\rho_m)$ (10⁻⁴ K⁻¹)
Zn	6.58	9.8	9.5–11.1	9.94	1.5

[a] 'Range' contains the extremes of modern experimental values for Λ which Steinberg (1974) considers reasonable.
Data for ρ_m, Λ, and Range from Steinberg (1974).
According to Hiemstra et al.'s experimental data (1977): Ca, $\rho_m = 1.378$, $\Lambda = 1.37$; Sr, $\rho_m = 2.351$, $-\Lambda = 2.67$; Ba, $\rho_m = 3.343$, $-\Lambda = 2.99$.

TABLE 3.2. Specific volumes, $v(\equiv \rho^{-1})$, of liquid high-melting-point metals.

Metal	a	v_l/v_0 b	c	Temperature range (K)	v_m (10⁻⁴ m³ kg⁻¹)	ρ_m (10³ kg m⁻³)
Ti	1.093	15.75	5.671	1943–5100	2.407	4.155
V	1.098	9.595	8.943	2175–6600	1.797	5.565
Mo	1.1199	9.912	2.194	2890–7000	1.099	9.099
Pd	–	–	–	–	0.972	10.29
W	1.18	6.20	32.3	3680–7500	0.611	16.37

The subscripts l, 0, and m refer to liquid, room temperature, and melting point, respectively.
$v_l/v_0 = a + b \times 10^{-5} (T - T_m) + c \times 10^{-9}(T - T_m)^2$.
For the calculation of v_l from v_l/v_0 handbook data for v_0 were used.
Data are taken from Seydel and Kitzel (1979).

TABLE 3.3. Estimated data for ρ_m and Λ of high-melting-point metals in liquid state

Metal	ρ_m (10³ kg m⁻³)	$-\Lambda$ (10⁻¹ kg m⁻³ K⁻¹)	V_m (10⁻⁶ m³ g-atom⁻¹)	α (10⁻⁴ K⁻¹)
Hf	12.0	6.3	14.9	0.53
Ir	20.0	–	9.61	–
Mo	9.35	5.0	10.3	0.53
Nb	7.83	4.0	11.9	0.51
Os	20.1	–	9.46	–
Re	18.7	8.2	9.96	0.44
Rh	11.1	–	9.27	–
Ru	10.9	–	9.27	–
Ta	15.0	6.9	12.1	0.46
Th	10.5	5.2	22.1	0.50
W	17.5	7.9	10.5	0.45
Zr	5.93	3.2	15.4	0.54

Data from Allen (1963, 1972a), and Steinberg (1974).

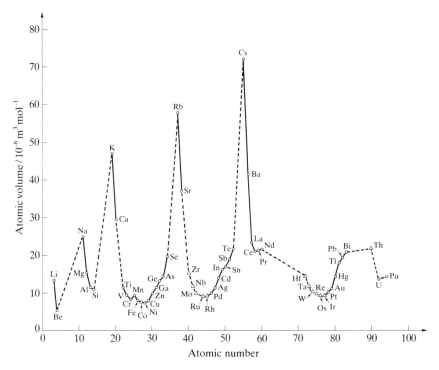

FIG. 3.17. The periodicity of atomic volumes of liquid metals at their melting points. △ estimated values (Allen 1963; Steinberg 1974).

range, i.e. from the melting point up to a certain ratio of the reduced temperature, some curvatures would have been probable towards the critical point temperatures.

3.5.2. Experimental densities of liquid binary alloys

The atomic volumes for most liquid binary alloys are actually linear functions of composition. While many liquid alloys show small deviations from this relationship, maximum values of V^E generally lie within ± 3 per cent, and more usually within ± 2 per cent. As a result, the atomic volumes of liquid binary alloys may be approximately evaluated by assuming additivity of component atomic volumes. For a multi-component system, atomic volumes may be estimated by

$$V \approx \Sigma \, x_i V_i \tag{3.21}$$

Some exceptions are the compound-forming alloys such as iron–silicon, mercury–sodium, mercury–potassium, sodium–lead, sodium–bismuth, aluminium–copper, magnesium–lead, sodium–indium (Crawley 1974; Marcus 1977; Kubaschewski and Alcock 1979c). These systems show large negative V^E

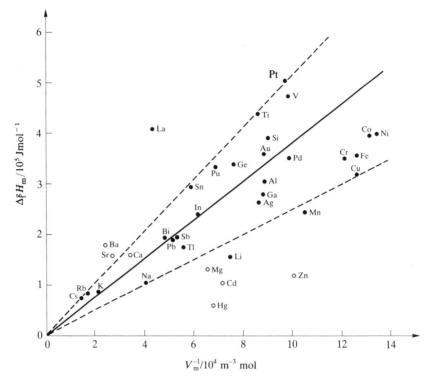

FIG. 3.18. Heat of vaporization (cohesive energy) as a function of atomic volume for various liquid metals at their melting points. ○ Group II metals. The broken lines denote ± 35 per cent error band.

values which are of the order of − 6 to − 27 per cent. Eutectic density minima and rapid changes in density with composition have been reported in a few systems. However, more careful investigations are necessary before the validity of these density anomalies is established. These include the discontinuities in temperature variations of liquid iron.

Recently, Morita, Iida, and Matsumoto (1985) have obtained accurate density data for liquid mercury–indium alloys, using a dilatometer with a capillary neck. In Figs. 3.19–3.21 their results are compared with Predel and Emam's data (1969). As can be seen from Fig. 3.19, the isothermal densities by Morita *et al.* (1985) and Predel and Emam (1979) both show positive deviations from the additivity of densities, but the maximum values of the deviations are 1.2 per cent and 4 per cent, respectively. Furthermore, the values for Λ also indicate rather large discrepancies. Agreement between them is not satisfactory, though the density data for low-melting-point systems are considered to be most reliable.

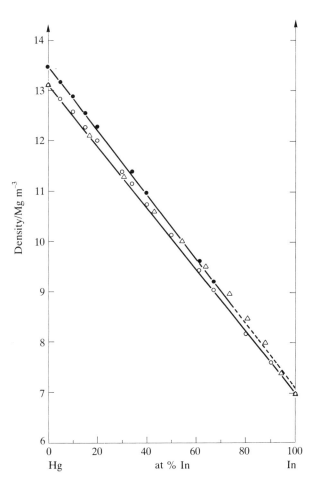

FIG. 3.19. Isothermal densities of liquid mercury–indium alloys at 323 K and 473 K (after Morita, Iida, and Matsumoto 1985). ● 323 K, ○ 473 K (Morita *et al.*); △473 K (Predel and Emam).

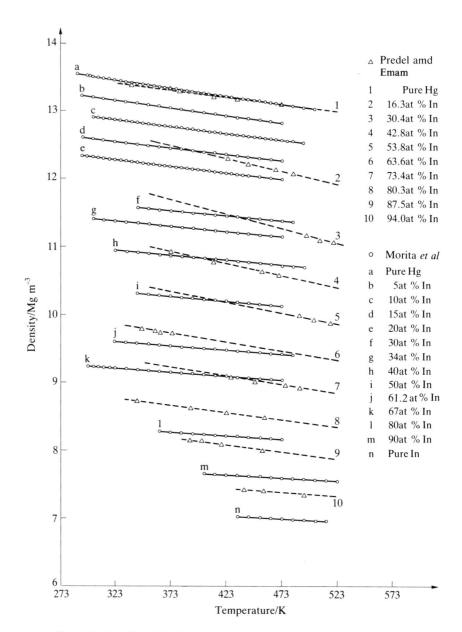

FIG. 3.20. Densities of liquid mercury–indium alloys (after Morita *et al.* 1985).

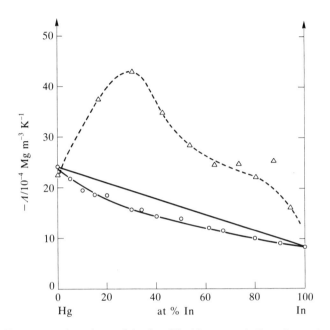

FIG. 3.21. Temperature dependence of density of liquid mercury–indium alloys (after Morita *et al*. 1985). ○ Morita *et al*., △ Predel and Emam.

Derivation of eqn (3.12) (Iida and Morita 1978)

$$\alpha_m U_c \approx A, \text{ or } -\Lambda U_c \approx A\rho_m. \tag{1}$$

substituting eqn (4.45) into the above equation, we have

$$\Lambda \approx -\frac{2A\xi\rho_m}{U_m^2 M} = \frac{2A\xi M^{1/2}}{U_m^2 V_m M^{1/2}}. \tag{2}$$

Equation (4.48) states that the values of $(U_m^2 V_m M^{1/2})$ are approximately constant for all liquid metals. Consequently, we have

$$\Lambda \approx -B\xi M^{1/2}, \ (B = 2A/U_m^2 V_m M^{1/2}). \tag{3}$$

The numerical factor of 2.8×10^{-5}, (i.e. the value of B), was determined so as to best fit the experimental data as indicated in Fig. 3.14.

4

THERMODYNAMIC PROPERTIES

4.1. INTRODUCTION

This chapter deals with some of the more important thermodynamic properties of liquid metals. These comprise vapour pressure, heat capacity, and sound velocity. Theoretical and empirical equations describing these thermodynamic properties are therefore presented, together with experimental data. As the experimental methods associated with these data are well documented in other textbooks, such as Kubaschewski and Alcock (1979*f*), they are not reported here.

4.2. VAPOUR PRESSURE OF PURE LIQUID METALS

4.2.1. Introduction

The vapour pressure of a liquid metal is an important property which is related to the liquid's cohesive or binding energy. From an engineering standpoint, a knowledge of a liquid metal's vapour pressure is of obvious importance. For instance, it is sometimes of technological importance that a pure metal's loss by evaporation must be minimized during processing operations. Alternatively, rates of evaporation of metallic elements or impurities may be utilized in the pyrometallurgical production of mercury, cadmium, zinc, and magnesium, and in refining processes. Since the equilibrium vapour pressure of a condensed phase varies exponentially with temperature, liquid metals tend to evaporate more rapidly with rising temperature.

From the standpoint of thermodynamics, the application of the Clausius–Clapeyron relation to equilibrium between a condensed liquid phase and a gaseous phase provides an expression for changes in vapour pressure with temperature. However, the thermodynamic equation contains an integration constant which has to be determined experimentally.

The Sackur–Tetrode equation, derived from statistical mechanics, can also give the equilibrium vapour pressure of a condensed phase and contains no undetermined constant.

4.2.2. Theoretical equations for vapour pressure

4.2.2.1. Thermodynamic equation

For two phases α and β in equilibrium with each other, the following relation exists between temperature and equilibrium vapour pressure P:

$$\frac{dP}{dT} = \frac{H^\beta - H^\alpha}{T(V^\beta - V^\alpha)} = \frac{\Delta_\alpha^\beta H}{T\Delta_\alpha^\beta V} \tag{4.1}$$

This is well known as the Clausius–Clapeyron equation. Let us now consider equilibrium between a liquid and a gaseous phase. Using the superscripts l for the liquid and g for the gas, we see from eqn (4.1) that the equilibrium pressure varies with equilibrium temperature according to

$$\frac{dP}{dT} = \frac{H^g - H^l}{T(V^g - V^l)} = \frac{\Delta_l^g H}{T\Delta_l^g V} \tag{4.2}$$

This rigorous thermodynamic equation can be simplified using the following approximations. First, we can assume the gas is ideal and neglect the second virial coefficient. Second, we can neglect the volume of liquid compared with that of the gas: $V^g \gg V^l$. For one mole of gas, we then have

$$V^g - V^l \approx V^g \approx \frac{RT}{P} \tag{4.3}$$

Substitution of eqn (4.3) in eqn (4.2) leads to

$$\frac{1}{P}\frac{dP}{dT} = \frac{d(\ln P)}{dT} = \frac{\Delta_l^g H}{RT^2} \tag{4.4}$$

or

$$\frac{d(\ln P)}{d(1/T)} = -\frac{\Delta_l^g H}{R} \tag{4.5}$$

From Kirchhoff's law, we obtain

$$\Delta_l^g H = H_0 + \int_0^T \Delta_l^g C_p \, dT \tag{4.6}$$

where H_0 is the enthalpy of evaporation at 0K, and $\Delta_l^g C_p = C_p^g - C_p^l$. On combining this equation with the Clausius–Clapeyron eqn (4.4) and integrating, we then have

$$\ln P = -\frac{H_0}{RT} + \int_0^T \frac{dT}{RT^2} \int_0^T \Delta_l^g C_p \, dT + i \tag{4.7}^\dagger$$

where i is the constant of integration (the so-called vapour pressure constant),

† $\int_0^T \frac{dT}{T^2} \int_0^T C_p \, dT = \int_0^T \frac{C_p}{T} \, dT - \frac{1}{T}\int_0^T C_p \, dT = \int_0^T C_p \, d(\ln T) - \frac{1}{T}\int_0^T C_p \, dT$

which must be determined experimentally. Assuming once more that the gas is ideal: $C_p^g = C_V^g + R = 5R/2$, and

$$\ln P = -\frac{H_0}{RT} + \frac{5}{2}\ln T - \frac{1}{R}\int_0^T C_p^l d(\ln T) + \frac{1}{RT}\int_0^T C_p^l dT + i \qquad (4.8)$$

To evaluate vapour pressures using eqn (4.8), it is clearly necessary to know the integration constant i and the temperature dependence of heat capacity $\Delta_l^g C_p$. Since $\Delta_l^g H$ or $\Delta_l^g C_p$ varies slowly with temperature, we can approximate the change $(\Delta_l^g H)$ with the following simple algebraic expression for the temperature variation of $\Delta_l^g H$:

$$\Delta_l^g H = A + BT + CT^2 \qquad (4.9)$$

where A, B, and C are constants. On substituting this empirical equation into eqn (4.4) and integrating,

$$\ln P = -\frac{A}{RT} + \frac{B}{R}\ln T + \frac{C}{R}T + D \qquad (4.10)$$

where D is the constant of integration.

Taking the enthalpy of evaporation as constant, eqn (4.4) can be integrated immediately, to give

$$\ln P = -\frac{\Delta_l^g H}{RT} + \ln A \qquad (4.11)$$

or

$$P = A \exp\left(-\frac{\Delta_l^g H}{RT}\right) \qquad (4.12)$$

where $\ln A$ is a constant of integration. Equation (4.12) can provide approximate values for equilibrium vapour pressures over a wide range of temperature. The relationship shows that vapour pressures should increase exponentially with temperature and that their temperature sensitivities depend on their $\Delta_l^g H$ values.

4.2.2.2. Statistical mechanical equation

According to the theory of statistical mechanics, the entropy of a perfect gas S^g is given by

$$S^g = \frac{5}{2}Nk\ln T - Nk\ln P + Nk\left(\frac{5}{2} + i\right) \qquad (4.13)$$

$$i = \ln\left\{\frac{(2\pi m)^{3/2} k^{5/2}}{h^3}\right\}$$

where N is the number of atoms (monatomic molecules), m is the mass of a single atom, k is Boltzmann's constant, h is Planck's constant, and i is the so-called chemical constant. This relation was derived independently by Sackur and Tetrode, and is known as the Sackur–Tetrode equation. The only approximation made in deriving the Sackur–Tetrode eqn (4.13) is to treat the vapour as being an ideal gas with a constant heat capacity. This is generally a good approximation.

Let us now consider a liquid and its vapour in equilibrium at temperature T. If S^g and S^l are molar entropies of gas and liquid, we have

$$S^g - S^l = \frac{\Delta_l^g H}{T} \tag{4.14}$$

Where $\Delta_l^g H$ is the molar enthalpy of evaporation of liquid at temperature T. On combining this equation with eqn (4.13) and solving for $\ln P$, we obtain

$$\ln P = -\frac{\Delta_l^g H}{RT} + \frac{5}{2}\ln T + \frac{5}{2} + \ln\left\{\frac{(2\pi m)^{3/2}k^{5/2}}{h^3}\right\} - \frac{S^l}{R} \tag{4.15}$$

This equation is also often called the Sackur–Tetrode equation. The entropy of liquid S^l appearing in eqn (4.15) can be calculated from measured heat capacities using the relation:

$$S^l = \int_0^T \frac{C_p^l}{T}\,dT = \int_0^T C_p^l\,d(\ln T) \tag{4.16}$$

One should note that eqn (4.15) contains no undetermined constants. Values calculated with this equation give excellent agreement with experimental data (e.g. Fowler and Guggenheim 1965).

4.2.3. Empirical equation for vapour pressure

It follows from eqn (4.5) that if we plot $\ln P$ against $1/T$, the curve so obtained should, at each point, have a slope equal to $\Delta_l^g H/R$. Figure 4.1 shows the plots of $\ln P$ against $1/T$ for liquid mercury and sodium. Actually, in treating experimental systems, one obtains straighter lines than might have been expected in view of the wide range of temperatures seen in Fig. 4.1. The explanation is that the variation in $\Delta_l^g H$ with respect to temperature is relatively small, and that those variations may be partly compensated by a certain degree of non-ideality in the gas.

Saturated vapour pressures of metals have frequently been represented by empirical formulae having the same form as eqn (4.11), viz. $\ln P = a - b/T$, where a and b are constants. The form of eqn (4.11) provides a good approximation, and is useful for interpolation between tabulated values. Closer fits are obtained when a further term is added: $\ln P = a - b/T + c\ln T$.

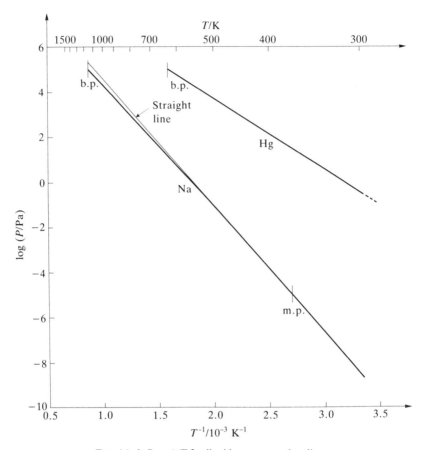

FIG. 4.1. ln P vs. $1/T$ for liquid mercury and sodium.

Theoretical considerations for this type of formula have already been mentioned. To a very good approximation, $\Delta_l^g H$ can be expressed by an equation of the form

$$\Delta_l^g H = \Delta_l^g H_m - K_e(T - T_m) \tag{4.17}$$

or

$$\left(\frac{\partial \Delta_l^g H}{\partial T}\right)_p = -K_e \tag{4.18}$$

where $T_m \leqq T \leqq T_b$. Using the data provided in Table 1.2, calculated values of K_e are given in Table 4.1. Considering experimental errors in the data

TABLE 4.1. *Temperature dependence of $\Delta_l^g H$ for various metallic elements*

Metal	$K_e{}^a$ $(\mathrm{J\,mol^{-1}\,K^{-1}})$	Metal	$K_e{}^a$ $(\mathrm{J\,mol^{-1}\,K^{-1}})$
Li	7.8	Sr	10.3
Na	9.9	Y	16.3
Mg	11.0	Mo	4.9
Al	8.4	Ag	6.7
Si	4.8	Cd	9.0
K	11.9	In	3.7
Ca	10.9	Sb	28.7
Ti	7.4	Cs	15.1
Cr	10.8	La	1.2
Mn	31.9	Sm	24.7
Fe	11.1	Ir	8.2
Ni	17.0	Pt	14.6
Cu	7.4	Au	7.9
Zn	10.3	Hg	5.6
Ga	4.2	Tl	6.1
Ge	6.4	Pb	7.8
Rb	15.0	Bi	10.6

$$^a\ K_e \equiv \frac{\Delta_l^g H_m - \Delta_l^g H_b}{T_b - T_m}$$

(experimental errors in the values of vapour pressure are of the order of 10–20 per cent), values for the temperature dependence of $\Delta_l^g H$ are of the order of 5–15 $(\mathrm{J\,mol^{-1}\,K^{-1}})$ for most liquid metals.

The temperature at which the vapour pressure of a saturated vapour is equal to one atmosphere $(P = 1 \text{ atm} = 1.01325 \times 10^5 \text{ Pa})$ is also known as the normal boiling point of a liquid, T_b. We then have from eqn (4.11) that

$$\frac{\Delta_l^g H_b}{RT_b} = \ln A \tag{4.19}$$

This relationship is known as Trouton's rule (see Subsection 1.2.2). $\ln A$ is a constant that is approximately the same for all metals.

To obtain enthalpies of evaporation $\Delta_l^g H$, or of sublimation $\Delta_s^g H$, plots of $\ln P$ against $1/T$ can be constructed by measuring P at several known temperatures. The slope of such a plot is, from eqn (4.5),

$$\text{Slope} = -\frac{\Delta_l^g H}{R} = \frac{\mathrm{d}(\ln P)}{\mathrm{d}(1/T)} = -\frac{T^2}{P}\frac{\mathrm{d}P}{\mathrm{d}T} \tag{4.20}$$

When eqn (4.2) is combined with the above relation, we have

$$\text{Slope} = -\frac{T\,\Delta_l^g H}{P\,\Delta_l^g V} \tag{4.21}$$

4.2.4. Experimental data for vapour pressure

Experimental values for the equilibrium vapour pressures of various elements are shown graphically in Figs. 4.2(a)–4.2(g). Their governing equations are summarized in Table 4.2.

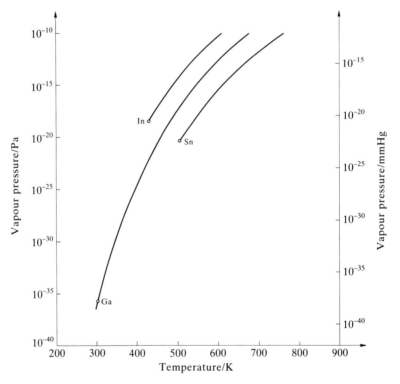

FIG. 4.2(a). Equilibrium vapour pressures of liquid metals as a function of temperature. ○ melting point.

Most normal metals are monatomic in the vapour form. However, the vapour species of semi-metals and non-metals (e.g. phosphorus, sulphur) are often polymerized. For example, bismuth vapour contains both Bi and Bi_2 gas species. For most metals, even for copper, silver, and gold, small amounts of dimer gas molecules have been observed.

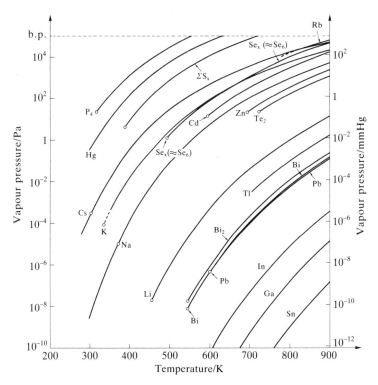

FIG. 4.2(b). Equilibrium vapour pressures of liquid metals as a function of temperature. ○ melting point.

4.3. HEAT CAPACITY OF PURE LIQUID METALS

4.3.1. Introduction

The heat capacity C of a substance is equal to the quantity δq added to, or withdrawn from, the substance, divided by the resultant change in temperature, dT, of that substance: $C = \delta q/dT$. Heat capacity is one of the most fundamental thermodynamic properties of matter. Calculations of changes in enthalpy or entropy of a substance with temperature require a knowledge of the temperature dependence of heat capacity.

Investigations of the heat capacities of solid elements have been made for many years. Dulong and Petit in 1819 proposed the empirical law that most solid metals have values of heat capacity lying around $26\,\mathrm{J\,mol^{-1}\,K^{-1}}$ ($6.2\,\mathrm{cal\,mol^{-1}\,K^{-1}}$) under standard conditions of pressure (1 atm) at room temperature. The theoretical work of Einstein, Debye, and others on the heat

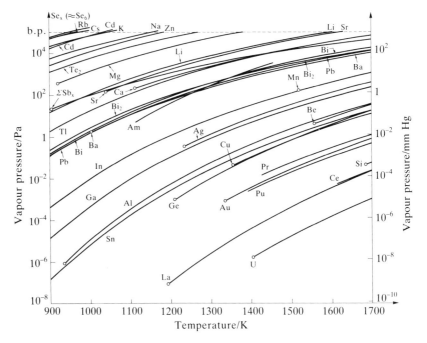

FIG 4.2(c). Equilibrium vapour pressures of liquid metals as a function of temperature ○ melting
point.

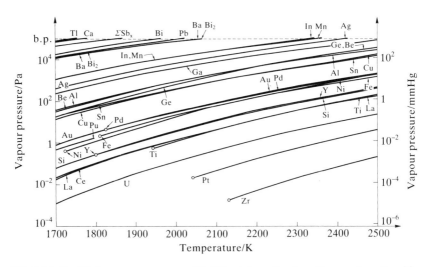

FIG. 4.2(d). Equilibrium vapour pressures of liquid metals as a function of temperature. ○ melting
point.

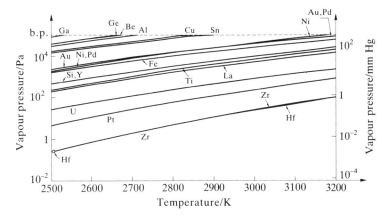

FIG. 4.2(e). Equilibrium vapour pressures of liquid metals as a function of temperature.

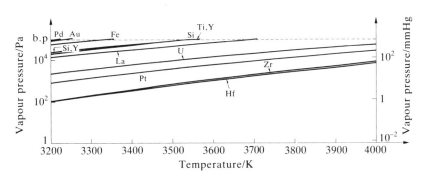

FIG. 4.2(f). Equilibrium vapour pressures of liquid metals as a function of temperature.

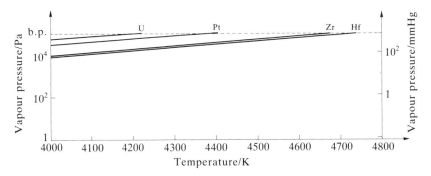

FIG. 4.2(g). Equilibrium vapour pressures of liquid metals as a function of temperature. (Figures 4.1 and 4.2(a)–4.2(g) are drawn using the data of Kubaschewski and Alcock 1979d.)

TABLE 4.2. *Experimental values for saturated vapour pressure of liquid metals, phosphorus and sulphur.* ($log = log_{10}$; 1 mm Hg = 133.322 Pa)

Element		Equation for vapour pressure (mmHg)	Temperature range (K)
Lithium	Li	$\log P = -8415/T - 1.0 \log T + 11.34$	m.p.–b.p.
Beryllium	Be	$\log P = -17000/T - 0.775 \log T + 11.90$	1557–2670
Sodium	Na	$\log P = -5780/T - 1.18 \log T + 11.50$	298–b.p.
Magnesium	Mg	$\log P = -7550/T - 1.41 \log T + 12.79$	m.p.–b.p.
Aluminium	Al	$\log P = -16380/T - 1.0 \log T + 12.32$	m.p.–b.p.
Silicon	Si	$\log P = -20900/T - 0.565 \log T + 10.78$	m.p.–b.p.
Potassium	K	$\log P = -4770/T - 1.37 \log T + 11.58$	350–1050
Calcium	Ca	$\log P = -8920/T - 1.39 \log T + 12.45$	m.p.–b.p.
Titanium	Ti	$\log P = -23200/T - 0.66 \log T + 11.74$	m.p.–b.p.
Manganese	Mn	$\log P = -14520/T - 3.02 \log T + 19.24$	m.p.–b.p.
Iron	Fe	$\log P = -19710/T - 1.27 \log T + 13.27$	m.p.–b.p.
Nickel	Ni	$\log P = -22400/T - 2.01 \log T + 16.95$	m.p.–b.p.
Copper	Cu	$\log P = -17520/T - 1.21 \log T + 13.21$	m.p.–b.p.
Zinc	Zn	$\log P = -6620/T - 1.255 \log T + 12.34$	m.p.–b.p.
Gallium	Ga	$\log P = -14330/T - 0.844 \log T + 11.42$	298–b.p.
Germanium	Ge	$\log P = -18700/T - 1.16 \log T + 12.87$	m.p.–b.p.
Selenium (mainly Se_6)	Se_x	$\log P = -4990/T + 8.09$	m.p.–b.p.
Rubidium	Rb	$\log P = -4688/T - 1.76 \log T + 13.07$	813–1258
Strontium	Sr	$\log P = -9000/T - 1.31 \log T + 12.63$	m.p.–b.p.
Yttrium	Y	$\log P = -22280/ - 1.97 \log T + 16.13$	m.p.–b.p.
Zirconium	Zr	$\log P = -30300/T + 9.38$	m.p.–b.p.
Palladium	Pd	$\log P = -17500/T + 1.0 \log T + 4.81$	m.p.–b.p.
Silver	Ag	$\log P = -14400/T - 0.85 \log T + 11.70$	m.p.–b.p.
Cadmium	Cd	$\log P = -5819/T - 1.257 \log T + 12.287$	594–1050
Indium	In	$\log P = -12580/T - 0.45 \log T + 9.79$	m.p.–b.p.
Tin	Sn	$\log P = -15500/T + 8.23$	505–b.p.
Antimony (total pressure)	Sb_x	$\log P = -6500/T + 6.37$	m.p.–b.p.
Tellurium	Te_2	$\log P = -7830/T - 4.27 \log T + 22.29$	m.p.–b.p.
Caesium	Cs	$\log P = -4075/T - 1.45 \log T + 11.38$	280–1000
Barium	Ba	$\log P = -9340/T + 7.42$	m.p.–
Lanthanum	La	$\log P = -21530/T - 0.33 \log T + 9.89$	m.p.–b.p.
Cerium	Ce	$\log P = -20304/T + 8.207$	1611–2038
Praseodymium	Pr	$\log P = -17190/T + 8.10$	1425–1692
Hafnium	Hf	$\log P = -29830/T + 9.20$	m.p.–b.p.
Platinum	Pt	$\log P = -27890/T - 1.77 \log T + 15.71$	m.p.–b.p.
Gold	Au	$\log P = -19280/T - 1.01 \log T + 12.38$	m.p.–b.p.
Mercury	Hg	$\log P = -3305/T - 0.795 \log T + 10.355$	298–b.p.
Thallium	Tl	$\log P = -9300/T - 0.892 \log T + 11.10$	700–1800
Lead	Pb	$\log P = -10130/T - 0.985 \log T + 11.16$	m.p.–b.p.
Bismuth	Bi	$\log P = -10400/T - 1.26 \log T + 12.35$	m.p.–b.p.
Bismuth	Bi_2	$\log P = -10730/T - 3.2 \log T + 18.1$	m.p.–b.p.
Uranium	U	$\log P = -24090/T - 1.26 \log T + 13.20$	m.p.–b.p.
Plutonium	Pu	$\log P = -17590/T + 7.90$	1390–1793
Americium	Am	$\log P = -13700/T - 1.0 \log T + 13.97$	1103–1453
Phosphorus	P_4	$\log P = -2740/T + 7.84$	m.p.–b.p.
Sulphur (total pressure)	S_x	$\log P = -4830/T - 5.0 \log T + 23.88$	m.p.–b.p.

Data from Kubaschewski and Alcock (1979*d*). Recently, vapour pressure equations have been presented by Alcock *et al.* (see the footnote of the end of this Chapter).

capacity of a solid element is well known. The values of the heat capacity of the solid calculated by the Einstein and the Debye models, which are based on statistical thermodynamics and quantum mechanics, are in moderately good agreement with those measured.

Clearly, thermodynamics gives no information about the numerical values of heat capacities nor of their temperature dependence. Various theoretical approaches have been made concerning the heat capacities of liquids. None has been sucessful, because the motion of atoms in the liquid state is extremely complex. Furthermore, no empirical relationship has been successful in providing reliable estimates of liquid metals' heat capacities. Accurate experimental data for the heat capacities of liquid metals and alloys are, therefore, highly desirable.

The values of the heat capacities of liquid alloys have often been estimated by proportional addition of the heat capacities of the constituent elements, i.e. the Neumann–Kopp law.

4.3.2. Heat capacity at constant pressure

For a given system, the value of the heat capacity depends upon the constraints imposed on the system. The two cases of most common interest are for conditions of (a) constant volume, and (b) constant pressure. These two heat capacities are defined by the equations

$$C_V = \frac{\delta q_V}{dT} = \left(\frac{\partial U}{\partial T}\right)_V \tag{4.22}$$

$$C_p = \frac{\delta q_p}{dT} = \left(\frac{\partial H}{\partial T}\right)_p \tag{4.23}$$

where δq represents an infinitessimal quantity of heat added to, or withdrawn from, a system, dT is the resulting infinitessimal temperature change, U is the internal energy, H is the enthalpy, and subscripts V or p denote conditions of constant volume or constant pressure, respectively.

C_p and C_V are interrelated according to the following thermodynamic relation:

$$C_p - C_V = \frac{\alpha^2 V T}{\kappa_T} \tag{4.24}$$

where κ_T represents isothermal compressibility. Equation (4.24) is useful since, if we are in possession of any five of these measurable quantities, the sixth can be deduced. Most metallurgical and chemical processes are carried out at constant pressure. Under constant pressure conditions and with no work other than reversible work of volume expansion or contraction,

$$\delta q_p = dH = C_p \, dT \tag{4.25}$$

On integrating eqn (4.25) between states (T_2, P) and (T_1, P), we have

$$\Delta H = H(T_2, P) - H(T_1, P) = \int_{T_1}^{T_2} C_p \, dT \qquad (4.26)$$

According to the second law of thermodynamics, the corresponding increase in entropy is

$$dS = \frac{dH}{T} = \frac{C_p}{T} \, dT \qquad (4.27)$$

Again, on integrating, we have

$$\Delta S = S(T_2, P) - S(T_1, P) = \int_{T_1}^{T_2} C_p \, d(\ln T) \qquad (4.28)$$

ΔH and ΔS are called the enthalpy and entropy increments, respectively.

In general, heat capacity will vary with temperature. The value of ΔH is obtained by plotting C_p against T and evaluating the area under the curve; similarly the value of ΔS is calculated by plotting C_p against $\ln T$ and evaluating that area under the resulting curve.

4.3.3. Empirical representation of heat capacity

Accurate data for the heat capacities of liquid metals are not abundant. For most liquid metals the values of heat capacity C_p are assumed to be constant over relatively wide ranges of temperature, because there is insufficient experimental data available to allow the temperature dependence of C_p to be determined. In general, the influence of temperature on liquid metal heat capacities tends to be small.

Table 4.3 lists values of constant-pressure heat capacity for various liquid elements. At their melting points, most liquid metals have C_p values ranging between 30 and 40 J mol^{-1} K^{-1} (7–10 cal mol^{-1} K^{-1}).

Table 4.4 gives arithmetical representations of C_p for 10 liquid elements. Within a certain temperature range, C_p, like that of solids, can be represented by the empirical equation:

$$C_p = a + bT + cT^{-2} + dT^2 \qquad (4.29)$$

in which a, b, c and d are constants.

Figures 4.3(a) and 4.3(b) show the variation of C_p with T for various liquid metals.

TABLE 4.3. *Molar heat capacities at constant pressure, C_p, for liquid elements*

Element	C_p ($J\,mol^{-1}\,K^{-1}$)	Temperature range (K)
Be	29.46	1560 (m.p.)–2800
B	31.38	2453 (m.p.)–3500
Mg	32.64	922 (m.p.)–1150
Al	31.8	933 (m.p.)–2400
Si	25.61	1685 (m.p.)–1873
P	26.32	317–870
Ca	29.29	1112 (m.p.)–1757
V	47.49	2175 (m.p.)–2600
Cr	39.33	2130 (m.p.)–
Mn	46.0	1517 (m.p.)–2333 (b.p.)
Fe	41.8	1809 (m.p.)–1873
Co	40.38	1768 (m.p.)–1900
Ni	38.49	1726 (m.p.)–2000
Cu	31.38	1356 (m.p.)–1600
Zn	31.38	693 (m.p.)–1200
Ge	27.61	1210 (m.p.)–1600
Se	35.1	493 (m.p.)–800
Y	39.79	1799 (m.p.)–2360
Ag	30.54	1234 (m.p.)–1600
Cd	29.71	594 (m.p.)–1100
Sb	31.38	904 (m.p.)–1300
Te	37.7	723 (m.p.)–873
Cs	31.88	303 (m.p.)–330
Ba	48.1	1002 (m.p.)–1125
Ce	37.70	1071 (m.p.)–1500
Pr	42.97	1204 (m.p.)–1500
Nd	48.79	1289 (m.p.)–1400
Ta	41.8	3250–4000
Au	29.29	1336 (m.p.)–1600
Tl	30.1	577 (m.p.)–1760
Th	46.0	2023 (m.p.)–3000

Data from Kubaschewski and Alcock (1979e).
Over the stated range of temperature, C_p is constant.

4.4. VELOCITY OF SOUND

4.4.1. Introduction

Only a few studies have been made of the velocity of sound in liquid metals, even though the velocity of sound is one of their most basic thermodynamic properties.

TABLE 4.4. *Molar heat capacities at constant pressure for liquid elements as a function of temperature*

Element	C_p (J mol^{-1} K^{-1})	Temperature range (K)
Li	$24.48 + 5.48 \times 10^{-3}T + 8.66 \times 10^5 T^{-2}$ $- 1.954 \times 10^{-6}T^2$	454 (m.p.)–1200
Na	$37.510 - 19.221 \times 10^{-3}T + 10.636 \times 10^{-6}T^2$	371 (m.p.)–1200
S	$449.8 - 959.8 \times 10^{-3}T - 208.87 \times 10^5 T^{-2}$ $+ 607.1 \times 10^{-6}T^2$	388–718
K	$37.179 - 19.12 \times 10^{-3}T + 12.30 \times 10^{-6}T^2$	336 (m.p.)–1037
Ga	$26.36 + 1.26 \times 10^5 T^{-2}$	303 (m.p.)–1200
In	$30.29 - 1.38 \times 10^{-3}T$	430 (m.p.)–800
Sn	$34.69 - 9.2 \times 10^{-3}T$	510–810
Hg	$30.38 - 11.46 \times 10^{-3}T + 10.155 \times 10^{-6}T^2$	298–630
Pb	$32.43 - 3.10 \times 10^{-3}T$	600 (m.p.)–1200
Bi	$20.00 + 6.15 \times 10^{-3}T + 21.13 \times 10^5 T^{-2}$	544 (m.p.)–820

Data from Kubaschewski and Alcock (1979e).

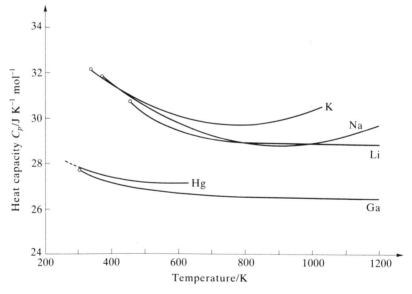

FIG. 4.3(a). Molar heat capacities at constant pressure for liquid metals as a function of temperature. ○ melting point.

The values of isentropic compressibility (sometimes referred to as adiabatic compressibility), isothermal compressibility, and constant-volume heat capacity of liquid metals can be determined readily using well-known thermodynamic formulae, provided that sound velocities can be measured, and that

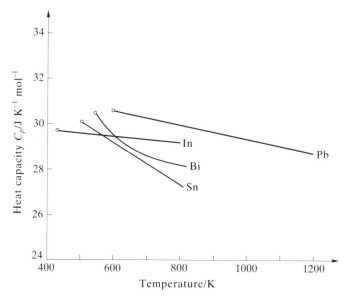

FIG. 4.3(b). Molar heat capacities at constant pressure for liquid metals as a function of temperature. ○ melting point (Figures 4.3(a) and 4.3(b) are drawn using the data of Kubaschewski and Alcock 1979e).

data for their constant-pressure heat capacities and thermal expansivities are available.

4.4.2. Thermodynamic relationship between velocity of sound and compressibility

The relationship between the velocity of sound U and the isentropic compressibility κ_S $(\equiv -V^{-1}(\partial V/\partial P)_S)$ is given by

$$U = \left(\frac{1}{\rho\kappa_S}\right)^{\frac{1}{2}} = \left(\frac{B_S}{\rho}\right)^{\frac{1}{2}} \tag{4.30}$$

in which ρ is the density of the medium and B_S is the isentropic bulk modulus. This equation forms the basis for determining κ_S or B_S from measurements of the velocity of sound.

The isothermal compressibility κ_T, the constant-volume heat capacity C_V, and the ratio of the heat capacities γ_h $(\equiv C_p/C_V)$ can be calculated through the thermodynamic relations

$$\frac{\kappa_T}{\kappa_S} = \frac{C_p}{C_V} \tag{4.31}$$

$$\frac{C_p}{C_V} = 1 + \frac{\alpha^2 V T}{\kappa_S C_p} \tag{4.32}$$

It is difficult to determine values for the isothermal compressibility and constant-volume heat capacity experimentally, but their values can be computed using these formulae together with sound velocity measurements.

4.4.2.1. Compressibility equations in terms of the pair distribution function and the structure factor

By working with a grand canonical ensemble and considering the fluctuations in the number of atoms in a given volume, isothermal compressibility can be formulated in terms of the pair distribution function as follows:

$$\kappa_T = \frac{1}{n_0 kT}\left[1 + n_0 \int_0^\infty \{g(r) - 1\}\, 4\pi r^2\, dr\right] \tag{4.33}$$

The isothermal compressibility is related to the structure factor $S(0)$, (i.e. the structure factor $S(Q)$ as $Q \to 0$, where $Q = 4\pi \sin\theta/\lambda$), as follows:

$$\kappa_T = \frac{1}{n_0 kT}S(0) \quad \left(\lim_{Q \to 0} S(Q) = S(0)\right) \tag{4.34}$$

At present, the accuracy of experimental data in the low Q region is low and not necessarily satisfactory.

4.4.3. Theoretical equations for the velocity of sound in liquid metals

(1) According to Bohm and Staver's considerations of the 'Jellium Model', which assumes the presence of a free non-interacting electron gas, the velocity of sound in condensed metal phases can be expressed by (Bohm and Staver 1951; Bardeen and Pines 1955; Ascarelli 1968)

$$U = \left(\frac{2ZE_F}{3M}\right)^{\frac{1}{2}} \tag{4.35}$$

where Z is the number of valence electrons per atom, E_F is the Fermi energy and M is the metal's atomic weight.

Computed and measured sound velocities in liquid metals at their melting points are given in Table 4.5. The ratios of measured values to those calculated using equation (4.35) lie between 0.4 and 1.6.

TABLE 4.5. *Comparison of calculated and measured values of the velocity of sound in liquid metals at their melting points.*

Metal	$(U_m)_{exp}$ (m s^{-1})	$(U_m)_{calc}$ (m s^{-1}) BS	A	GM	$(U_m)_{exp}/(U_m)_{calc}$ BS	A	GM
Na	2531	2960	2500	–	0.86	1.01	–
Al	4688	8750	4900	–	0.54	0.96	–
K	1880	1810	1720	–	1.04	1.09	–
Cu	3485	2580	2700	3450	1.35	1.29	1.01
Zn	2836	4180	2610	1980	0.68	1.09	1.43
Ga	2873	5430	2850	3020	0.53	1.01	0.95
Rb	1260	1140	1103	–	1.11	1.14	–
Ag	2810	1740	1920	2550	1.61	1.46	1.10
Cd	2242	2850	1840	1610	0.79	1.22	1.39
In	2314	3760	2041	2260	0.62	1.13	1.02
Sn	2466	4630	2440	2430	0.53	1.01	1.01
Sb	1893	5340	3150	1900	0.35	0.60	1.00
Cs	967	880	890	–	1.10	1.09	–
Hg	1480	2100	1220	–	0.70	1.21	–
Tl	1665	2760	1580	1400	0.60	1.05	1.19
Pb	1826	3350	1900	–	0.55	0.96	–
Bi	1670	3940	2080	1430	0.42	0.80	1.17

The experimental data are from Table 4.7.
BS: Bohm and Staver (Ascarelli 1968).
A: Ascarelli (1968); C_p/C_V has been taken as equal to 1.15 for all metals.
GM: Gitis and Mikhailov (Beyer and Ring 1972).

(2) Ascarelli (1968) has proposed an equation for the velocity of sound in liquid metals based on a model of hard spheres immersed in a uniform background potential. His final expression for U is:

$$U = \left[\frac{1}{M} \frac{C_p}{C_V} kT \left\{ \frac{(1+2\eta)^2}{(1-\eta)^4} + \frac{2}{3} \frac{ZE_F}{kT} - A \left(\frac{V_m}{V} \right)^{\frac{1}{3}} \frac{4kT_m}{3} \frac{1}{kT} \right\} \right]^{\frac{1}{2}} \quad (4.36)$$

where η is the packing fraction, V_m is the volume of the system at the melting point T_m, V is the volume of the system, and A is defined by

$$A \equiv 10 + \tfrac{2}{5} ZE_F (T_m)/kT_m$$

At the melting point of a liquid metal, eqn (4.36) becomes

$$U_m = \left[\frac{1}{M} \frac{C_p}{C_V} kT_m \left\{ 27 + \frac{1}{5} \left(\frac{2}{3} \frac{ZE_F}{kT_m} \right) \right\} \right]^{\frac{1}{2}} \quad (4.37)$$

As can be seen from Table 4.5, the ratio of the observed values for U_m to those calculated using eqn (4.37) ranges from 0.6 to 1.5 for the 17 metallic elements listed. It is evident that Ascarelli's results represent an improvement over those of the Bohm and Staver model, particularly for polyvalent metals.

(3) Using arguments involving the statistical mechanical theory of liquids, Gitis and Mikhailov (1966, 1968) have derived another expression for the velocity of sound in terms of the cohesive energy U_c and the atomic weight M as follows:

$$U = \left(\frac{2U_c}{M} \right)^{\frac{1}{2}} \tag{4.38}$$

where U_c is given by

$$U_c = \frac{2\pi N_A^2}{V} \int_0^\infty g(r)\phi(r)r^2 \, dr$$

N_A being Avogadro's number, and V atomic volume.

As indicated in Table 4.5, values computed through the use of eqn (4.38) lie reasonably close to experimental data, with the exception of zinc and cadmium (IIB metals), and, to a lesser extent, thallium and bismuth†.

From eqn (2.13) the heat of evaporation per mole at constant pressure, i.e. the enthalpy of evaporation per mole $\Delta_l^g H$, is given by

$$\Delta_l^g H = RT - \frac{2\pi N_A^2}{V} \int_0^\infty g(r)\phi(r)r^2 \, dr = RT - U_c \approx -U_c \quad (RT \ll |U_c|)$$

4.4.3.1. Temperature dependence of the velocity of sound

Ascarelli (1968) has also provided calculated values for the temperature dependence of the velocity of sound. These are: for rubidium, -0.3; for zinc, -0.20 (low-temperature value), -0.6 (high-temperature value); for indium, -0.14; and for tin, $-0.15 \text{ m s}^{-1} \text{ K}^{-1}$, respectively. Considering the simplicity of the model, agreement with experimental data is good (see Table 4.7).

Gitis and Mikhailov (1966) have proposed the following correlation between the velocity of sound and a liquid metal's electrical conductivity σ_e:

$$\frac{1}{\sigma_e} \frac{d\sigma_e}{dT} = \frac{2}{U} \frac{dU}{dT} + \frac{2}{3\rho} \frac{d\rho}{dT} \tag{4.39}$$

On the basis of measured values for U and ρ, they determined temperature

---

† The U_c values of liquid metals can be obtained from data on enthalpies of evaporation.

coefficients of electrical conductivity $(1/\sigma_e)(d\sigma_e/dT)$ for eight liquid metals using eqn (4.39). In most cases, their calculations yielded results which were about 15 to 20 per cent lower than experimental data.

4.4.4. Semi-empirical and empirical equations for the velocity of sound in liquid metals

(1) According to Einstein (1911), a simple relationship exists between the velocity of sound in solids and the mean frequency of atomic vibration v. His relation can be written as

$$U \propto v V^{\frac{1}{3}} \tag{4.40}$$

in which V is the atomic volume. By using Lindemann's formula, (1.5), for the mean atomic frequency, we have

$$U_m \propto \left(\frac{RT_m}{M}\right)^{\frac{1}{2}} \tag{4.41}$$

where R is the gas constant. This relationship shows that the velocity of sound in solids at their melting points, U_m, is proportional to the square root of the metal's melting point and inversely proportional to the square root of the metal's atomic weight. The expression has been known for a long time.

Equation (4.41) can be expected to be roughly valid for liquid metals, owing to the similarity of their sound velocities with metals in their solid state. Figure 4.4 shows this relation for a number of liquid metals.

(2) As already mentioned, Lindemann's melting law provides only rough values for the average vibrational frequency of atoms. In the semi-empirical treatment of Iida and co-workers (Iida *et al.* 1974; Kasama *et al.* 1976) for the surface tension of liquid metals, it was proposed that Lindemann's formula needed correcting for the mean frequency of atoms in the liquid state. At the melting point, the mean atomic frequency of liquid metals can be expressed (see Subsection 5.4.(3)) in the form

$$v = \beta v_L = 3.1 \times 10^8 \beta \left(\frac{RT_m}{M V_m^{\frac{2}{3}}}\right)^{\frac{1}{2}} \tag{4.42}$$

in which v_L is the atomic frequency which is calculated from Lindemann's formula by eqn (1.5), and the correction factor β is given by

$$\beta = \frac{1.1 \times 10^3 V_m^{\frac{1}{3}}}{\alpha} \left(\frac{\gamma_m}{RT_m}\right)^{\frac{1}{2}}$$

where

$$\alpha = \left(\frac{\rho_m}{\rho_c}\right)^{\frac{1}{3}} - 1 \equiv \left(\frac{V_c}{V_m}\right)^{\frac{1}{3}} - 1$$

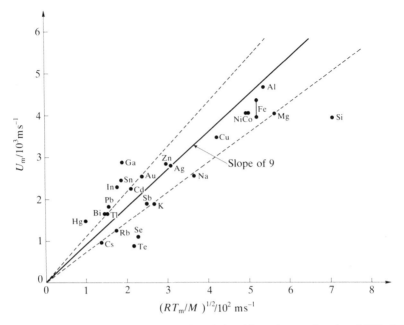

FIG. 4.4. The velocity of sound in liquid metals at their melting points as a function of $(RT_m/M)^{\frac{1}{2}}$. The broken lines denote ± 20 per cent error band.

and γ_m is the surface tension of the liquid metal at its melting point. The values of ρ_c or V_c have been determined experimentally for only a few metals. According to Young and Alder (1971), the atomic volume at the critical point, V_c, can be expressed in terms of the packing fraction at the melting point, η_m. From Young and Alder's consideration of V_c, which was based on the Van der Waals model, (though this model is not so suitable for metals), α is approximately equal to $(1.97\,\eta_m^{1/3} - 1)$ (Kasama et al. 1976). Combining eqn (4.42) for the mean atomic frequency with (4.40), we then obtain

$$U_m \propto \beta \left(\frac{RT_m}{M} \right)^{\frac{1}{2}} \tag{4.43}$$

Figure 4.5 indicates that this expression provides a better correlation for the various liquid metals for which data are available. As can be seen, sound velocities for 22 metallic elements, with the exception of silicon, lie within a ± 20 per cent error band. The slope of the line is 18 in SI units. This improvement over the former correlation (eqn (4.41)) is particularly remarkable for gallium, tin, indium, and mercury. (Values of η_m for selenium and tellurium are not yet available.) Values for β are given in Table 4.6.

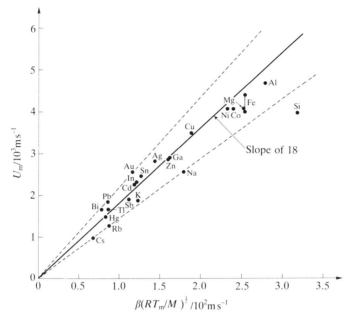

FIG. 4.5 The velocity of sound in liquid metals at their melting points as a function of $\beta(RT_m/M)^{\frac{1}{2}}$. The broken lines denote ± 20 per cent error band.

As mentioned in Section 2.3 the packing fraction at the melting point, η_m, is, to a good approximation, equal to 0.45 for all liquid metals. Consequently, α becomes approximately equal to 0.51 (i.e. $\alpha = 1.97\,\eta_m^{1/3} - 1$) for all liquid metals. We then have

$$U_m \propto \left(\frac{\gamma_m}{M}\right)^{\frac{1}{2}} V_m^{\frac{1}{3}} \tag{4.44}$$

Figure 4.6 shows this relationship. The graph gives good results for a number of liquid metals, having a slope of 3.8×10^4, as the constant of proportionality between sound velocity and the grouping $(\gamma_m/M)^{1/2} V_m^{1/3}$.

(3) As can be seen from Table 4.5, only ten liquid metals at their melting points have values for which both calculated and experimental data are available. Let us now assess the Gitis and Mikhailov eqn (4.38) for various metallic elements. In Fig. 4.7, measured sound velocities in liquid metals at their melting points are shown as a function of $(U_c/M)^{1/2}$. We see that the Gitis and Mikhailov relation is roughly true for a number of metals. However, the Group IIB metals (zinc, cadmium, and mercury), and Group VIB semi-metals and non-metal (selenium, tellurium, and sulphur), show considerable deviations from the line of proportionality. Further, the alkali metals (sodium, potassium, rubidium, and caesium) lie on their own straight line.

TABLE 4.6. *Values of β and ξ for various liquid elements*

Element	β	ξ
Na	0.49	0.69
Mg	0.45	1.51
Al	0.52	0.97
Si	0.38	0.57
S	–	0.11
K	0.50	0.79
Fe	0.48	1.24
Co	0.48	1.21
Ni	0.47	1.20
Cu	0.46	1.21
Zn	0.54	2.24
Ga	0.81	1.03
Se	–	0.24
Rb	0.47	0.80
Ag	0.47	1.60
Cd	0.56	2.72
In	0.68	1.29
Sn	0.64	1.23
Sb	0.40	1.12
Te	–	0.32
Cs	0.47	0.81
Au	0.49	1.80
Hg	0.84	3.58
Tl	0.55	1.64
Pb	0.55	1.83
Bi	0.54	1.50

From the microscopic point of view, sound velocity and compressibility are physical properties which are related to the curvature of the interatomic potential energy curve. Consequently, the velocity of sound in liquid metals should be linked to both repulsive and attractive energies between metal atoms. We may, therefore, propose the following expression for the velocity of sound in liquid metals:

$$U_m = \left(\frac{2\xi U_c}{M}\right)^{\frac{1}{2}} \tag{4.45}$$

wherein the correction factor ξ is some parameter related to the repulsive energy of the pair interaction potential. Values for ξ deduced on the basis of eqn (4.45) using experimental sound velocity data are listed in Table 4.6. One sees that ξ values vary between 0.1 and 3.6 for the 26 elements considered.

In Fig. 4.8, calculated values for ξ are plotted as a function of their respective atomic numbers. The figure shows that the values of ξ vary periodically with

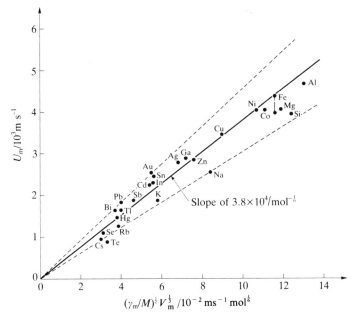

FIG. 4.6. The velocity of soumd in liquid metals at their melting points as a function of $(\gamma_m/M)^{\frac{1}{2}} V_m^{\frac{1}{3}}$. The broken lines denote ± 20 per cent error band.

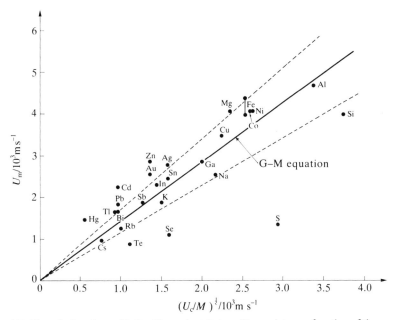

FIG. 4.7. The velocity of sound in liquid metals at their melting points as a function of the square root of (cohesive energy U_c divided by atomic weight M). The broken lines denote ± 20 per cent error band.

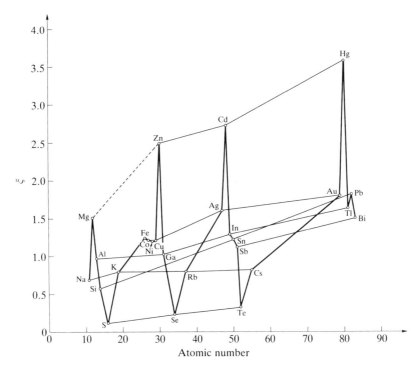

FIG. 4.8. The atomic periodicity in the values of ξ for various elements.

atomic number, the periodic Group II metals occupying the peaks, and the Group VI elements the valleys, of the curve. Restricting one's attention to only the normal metals, the periodic Group IA metals (i.e. the alkali metals) account for the smallest ξ values. This simple periodicity (Fig. 4.8) allows one to estimate values ξ and U_m for those metallic elements whose sound velocities have not yet been measured.

Matsuda and Hiwatari (1973) chose the following pair interaction $\phi(r)$ for condensed systems:

$$\phi(r) = \varepsilon \left(\frac{\sigma}{r}\right)^n - \alpha\delta \exp\left(-\delta r\right) \tag{4.46}$$

in which $\varepsilon > 0$, $\sigma > 0$, $n > 3$, $\alpha \geq 0$, $\delta > 0$. They computed values for the repulsive exponent n through the use of measured values of the bulk modulus at $0\,K$. A correlation between the correction factor ξ and the repulsive exponent n is shown for various liquid metals in Fig. 4.9. As can be seen, a simple linear relation of the form

$$\xi = 0.15\,n \tag{4.47}$$

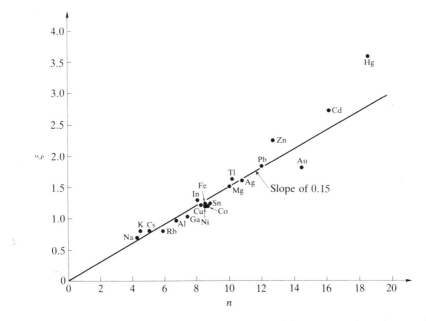

FIG. 4.9. Relation between the correction factor ξ and the repulsive exponent n for various metals.

holds. We can, therefore, consider that the correction factor ξ is a useful parameter representing the repulsive energy of the pair potential.

(4) According to Fig. 3.18, the cohesive energy U_c ($\approx \Delta_l^g H_m$) is roughly proportional to the reciprocal of a metal's atomic volume at its melting point V_m (except for Group II metals). On the other hand, the parameter ξ is approximately proportional to the square root of a metal's atomic weight, the Group II metals lying on their own straight line. Substituting these correlations into eqn (4.45), we have, therefore,

$$U_m = \frac{30}{M^{\frac{1}{4}} V_m^{\frac{1}{2}}} \tag{4.48}$$

Figure 4.10 shows a better than expected correlation between a liquid metal's sound velocity and $M^{-1/4} V_m^{-1/2}$. It may be that the effects of atomic volume and atomic weight on sound velocity compensate for each other.

(5) Through consideration of the interatomic forces between metal atoms, Wen-Po (1937) has deduced a simple relation between sound velocity, surface tension, and liquid density (see Taylor 1954):

$$U \propto \left(\frac{\gamma}{\rho}\right)^{\frac{2}{3}} \tag{4.49}$$

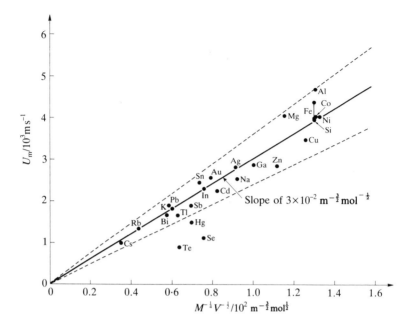

FIG. 4.10. The velocity of sound in liquid metals at their melting points as a function of $(M^{-\frac{1}{4}}V_m^{-\frac{1}{2}})$. The broken lines denote ± 20 per cent error band.

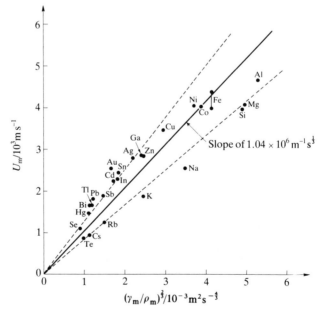

FIG. 4.11. The velocity of sound in liquid metals at their melting points as a function of $(\gamma_m/\rho_m)^{\frac{2}{3}}$ The broken lines denote ± 20 per cent error band.

TABLE 4.7. *Experimental values of the velocity of sound in liquid metals at their melting points, U_m, together with values of the temperature coefficient $-dU/dT$*

Element	T_m (K)	U_m (m s^{-1})	$-dU/dT$ (m s^{-1} K^{-1})	Ref.
Na	371	2531	0.54	(1)
Mg	923	4070[a]	0.62[a]	(2)
Al	933	4750	0.48	(3)
		4673, 4730	0.47, 0.16	(1)
		4688 $\pm$ 8	0.447	(4)
Si	1687	3977		(5)
S	388	1360[a]	See Fig.4.13	(6)
K	337	1880	0.53	(1)
Fe	1808	4400	$-$	(3)
		3983 $\pm$ 27	1.0021	(7)
Co	1765	4090[a]	0.51[a]	(8)
		4033 $\pm$ 20	0.5325	(7)
Ni	1728	4045		(5)
		4036 $\pm$ 5	0.3501	(7)
Cu	1356	3440	0.50	(3)
		3460	0.46	(1)
		3485 $\pm$ 8	0.524	(4)
Zn	693	2840	0.29	(3)
		2850, 2851	0.31, 0.40	(1)
		2836 $\pm$ 26		(9)
Ga	303	2950	0.28	(3)
		2873	0.27	(1)
Se	493	1100[a]	See Fig. 4.13	(6)
Rb	312	1260	0.4	(1)
Ag	1234	2770	0.47	(3)
		2710	0.41	(1)
		2810 $\pm$ 10	0.336	(4)
Cd	594	2235	0.40	(3)
		2256, 2242	0.29, 0.38	(1)
In	430	2314	0.28	(1)
Sn	505	2420	0.20	(3)
		2474	0.27	(1)
		2466 $\pm$ 10		(9)
Sb	904	1900	0.18[b]	(3)
		1893	-0.23	(1)
Te	723	889	-0.1	(1)
			(See Fig. 4.12)	
Cs	302	967	0.3	(1)
Au	1336	2560[a]	0.55[a]	(8)
Hg	234	1480	0.45	(1)
Tl	576	1665	0.23	(1)
Pb	601	1810	0.38	(3)
		1820	0.28	(1)
		1826 $\pm$ 12		(9)

TABLE 4.7. (*continued*)

Element	T_m (K)	U_m (m s^{-1})	$-dU/dT$ (m s^{-1} K^{-1})	Ref.
Bi	544	1620	0.21[b]	(3)
		1647	-0.1	(1)
		1670 ± 6		(4)

[a] These values were obtained from graphs.
[b] Non-linear temperature dependence exhibited.
Sources of data: (1) Beyer and Ring (1972); (2) Maier and Steeb (1973); (3) Filippov, Kazakov, and Pronin (1966); (4) Tsu, Suenaga, Takano, and Shiraishi (1982); (5) Sokolov, Katz, and Okorokov (1977); (6) Abowitz (1977); (7) Shiraishi and Tsu (1982); (8) Steeb and Bek (1976); (9) Tsu, Shiraishi, Takano, and Watanabe (1979).

Figure 4.11 shows that this relation is roughly valid for many liquid metals.

To the authors' knowledge, there are no experimental data on the velocities of sound for liquids such as calcium, titanium, vanadium, chromium, manganese, etc. Estimated values for these metals at their melting points have, therefore, been calculated using eqn (4.43), leading to the following predictions for U_m (m s^{-1}): calcium, 3540; titanium, 5230; vanadium, 5150; chromium, 4520; manganese, 3710.

4.4.5. Experimental data for the velocity of sound

Experimental data for the velocity of sound in liquid metals are not plentiful. However, the accuracy of the data is comparatively good compared with accuracies in other thermodynamic and physical properties of liquid metals. Nonetheless, in order to evaluate the other properties of liquid metals, for example, compressibility, using measured velocities of sound, even more accurate data are needed since $\kappa_s = \rho U^2$ and any errors are therefore approximately doubled.

Table 4.7 lists experimental results for the velocity of sound in liquid metals at their melting points. The experimental errors in the data are of the order of $\pm 0.5 - \pm 5$ per cent.

In general, the velocity of sound decreases linearly as metal temperature rises, with the temperature dependence dU/dT varying from -0.2 to -0.6 m s^{-1} K^{-1}. As can be seen from Table 4.7, there are larger discrepancies between some of the measured values of temperature gradients of sound velocity.

Several liquid metals exhibit non-linear variations in their velocity of sound with respect to temperature. As is apparent from Fig. 4.12, the velocity of sound in liquid tellurium first increases very rapidly and linearly with dU/dT

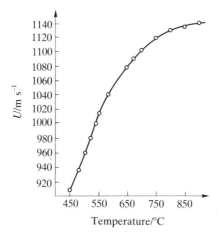

FIG. 4.12. The temperature dependence of the velocity of sound in liquid tellurium (after Gitis and Mikhailov 1966).

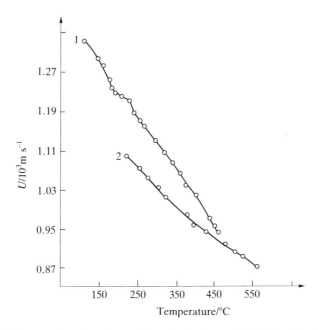

FIG. 4.13. The temperature dependence of the velocity of sound in liquid sulphur (curve 1) and selenium (curve 2) (after Gitis and Mikhailov 1967).

≈ 0.8 (m s^{-1} K^{-1}). It then slows markedly. Figure 4.13 shows experimental data for the temperature dependence of the velocity of sound in liquid sulphur and selenium, respectively (Gitis and Mikhailov 1967). These d$U/$dT values cease to be constant with increasing temperature. The phenomenon can be attributed to changes in liquid structure.

In closing, it is noted that few data are available on the velocity of sound in liquid alloys (Maier and Steeb 1973; Steeb and Bek 1976; Wilson 1965e).

We should also note that vapour pressure equations for the 65 metallic elements (298–2500 K) have been presented quite recently by Alcock *et al.* (Alcock, C. B., Itkin, V. P., and Horrigan, M. K. (1984). *Can. Met. Q.*, **23**, 309). In that paper, for each element, a precise four-term equation and a more practically-based equation, have been derived, through a fresh evaluation of thermochemical data for the condensed and gaseous elements. Their 'precise' equations reproduce the data to better than $\pm 1\%$, while their 'practical' equations provide better than $\pm 5\%$ accuracy.

5

SURFACE TENSION

5.1. INTRODUCTION

A knowledge of the surface tension of metals is essential to an understanding of metallurgical processes. In smelting and refining operations, the surface tension or interfacial tension is a dominating factor for phenomena such as gas absorption, nucleation of gas bubbles, nucleation and growth of non-metallic inclusions, and slag/metal reactions. Figure 5.1 provides an example of the important role surface tension can play in metallurgical mass transport phenomena. As can be seen, there is a simple relation between the surface tension of melts and apparent mass transfer coefficients. It will be appreciated that many other metallurgical technologies such as casting, brazing, sintering,

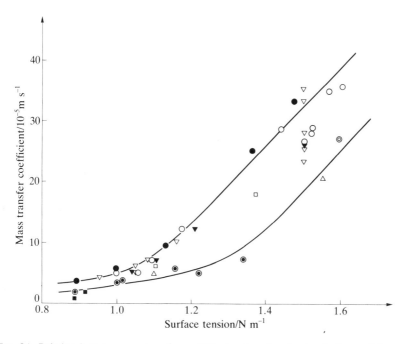

FIG. 5.1. Relation between mass transfer coefficient and surface tension in iron–sulphur and iron–oxygen melts (after Inouye and Choh 1968): Inouye and Choh, ○ Fe–O 1873 K, ● Fe–S 1873 K, ◉ Fe–S 1823 K, ◎ pure iron (0.007 per cent O) 1823 K; Elliott and Pehlke, ▽ Fe–O 1879 K, ▼ Fe–S 1879 K, △ Fe–O 1829 to 1834 K; Schenck, Frohberg, and Heinemann, □ Fe–O 1833 K, ■ Fe–S 1833 K.

zone melting, and fibre formation are greatly influenced by the role played by a liquid metal's surface tension.

Unfortunately, experimental and theoretical investigations of the nature and behaviour of surfaces or interfaces between phases are unsatisfactory. This situation results from the difficulty of precise experimental determinations of surface tensions or interfacial tensions. Furthermore, from a more fundamental viewpoint, we have, as yet, no experimental methods available for determining the structures of an interface, i.e. the transition region between two phases. Therefore, some assumptions are necessarily involved in any theoretical treatment of surface tension or interfacial tension.

5.2. EXPERIMENTAL MEASUREMENTS OF SURFACE TENSION

5.2.1. Characteristic features of experimental investigations on the surface tension of liquid metals

The plane of separation between two phases is known as an interface. In particular, an interface between a condensed phase and its own vapour or an inert gas is called a surface. The surface is in a higher energy state than the bulk liquid phase, because coordination among the atoms at the surface is incomplete.

Thermodynamically, the surface tension is defined as the surface free energy per unit area. In a dynamic sense, surface tension represents the work or energy required to create one unit of additional surface area at constant temperature. The dimensions of energy per unit area $J\ m^{-2}$ are equivalent to the dimensions of force per unit length $N\ m^{-1}$.

The surface tensions of most liquid metals have been determined experimentally. The exceptions are scandium, yttrium, technetium, and a majority of the rare earths and actinides. The accurate measurement of surface tension is not easy because of the difficulty of maintaining uncontaminated liquid metal or alloy surfaces at high temperatures. Similarly, some difficulties are involved in the adequate design of equipment for high temperatures. Table 5.1 and Figs. 5.2–5.4 respectively, give values of the surface tensions of liquid mercury, copper, and iron at their melting points. Mercury is the only liquid metal whose surface tension is accurately known. In other words, the surface tensions of liquid metals, particularly those with high melting points and/or those which are chemically reactive, have yet to be established.

In experimental determinations of the surface tension of liquid metals, the surface tension of mercury at room temperature is employed as an important standard for calibration purposes. As indicated in Fig. 5.2, however, the surface tension of mercury has taken us many years to establish. Even at the present time, the surface tension of liquid iron exhibits considerable scatter, as indicated in Fig. 5.4. As can be seen from Figs. 5.2–5.4, in general the surface

TABLE 5.1. *Surface tension values for liquid mercury at room temperature*

Investigator	Temp. (K)	γ (mN/m)	Method	Atm.	Year
Quincke	293	(542)	Pendant drop	H_2	1868
Sientopf	289	(410)	Sessile drop	air	1897
Schmidt	293	(435.5)	Oscillating jet	air	1912
Hagemann	293	472	Oscillating jet	vac.	1914
Palacios	293	(402)	Max. drop pres.[a]	vac.	1920
Harkins, Ewing	298	476	Drop weight		1920
Popesco	293	(436)	Sessile drop	vac.	1921
Richards, Boyer	293	(432)	Sessile drop	vac.	1921
Hogness	298	476	Max. drop pres.	H_2	1921
Iredale	298	472	Drop weight		1923
Iredale	298	(464)	Sessile drop		1924
Sauerwald, Drath	292	473	Max. bub. pres.[b]	H_2	1926
Oppenheimer	293	(437)	Capillary depre.[c]		1928
Bircumshaw	293	480	Drop weight	vac.	1928
Cook	298	(516)	Sessile drop		1929
Kernaghan	298	(435)	Sessile drop		1931
Brown	298	473	Drop weight	vac.	1932
Burdon	298	488	Sessile drop		1932
Bradley	298	(498)	Sessile drop		1934
Sauerwald *et al.*	293	(420)	Sessile drop	vac.	1935
Kernaghan	298	476	Sessile drop		1936
Semenchenko, Pokrovski	293	(410)	Max. drop pres.	vac.	1937
Didenko, Pokrovski	293	(455)	Max. drop pres.	vac.	1941
Kemball	298	484 ± 1.5	Sessile drop		1946
Pugachevich	295	(468)	Max. drop pres.	vac.	1951
Ziesing	298	484.9 ± 1.8	Sessile drop		1953
Bering, Ioileva	293	485.5 ± 1.0	Max. drop pres.	vac.	1953
Taylor	293	(454.7)	Max. bub. pres.	Ar	1954
Gratzianski, Rjabov	263	487	Max. drop pres.	vac.	1959
Bobyk	294	(350.5)	Max. bub. pres.		1960
Fessenko, Eremenko	293	475	Max. bub. pres.	He, H_2	1960
Korolkov	293	(500 ± 15)	Max. bub. pres.	Ar	1960
Nicholas *et al.*	298	483.5 ± 1.0	Sessile drop	vac.	1961
Timofejevicheva, Lasarev	295	(465)	Max. bub. pres.		1962
Olson, Johnson	298	485.1	Sessile drop		1963
Roberts	298	485.4 ± 1.2	Pendant drop		1964
Biery, Oblak	296–298	482.8 ± 9.7	Contact angle		1966
Yung	293	(465.4)			1968
Melik-Gajkazan *et al.*	293	484.6 ± 1.3	Pendant drop	vac.	1968
Roehlich *et al.*	298	480	Max. bub. pres.	Ar	1968
Schwaneke *et al.*	293	482.5 ± 3.0	Max. bub. pres.	vac.	1970
Lang	294.7	484.9 ± 0.3	Max. bub. pres.	Ar	1973

[a] Maximum drop pressure.
[b] Maximum bubble pressure.
[c] Capillary depression.
Parentheses indicate that the values are less reliable.
Note: The surface tension of pure liquid mercury is considered to have been established. The surface tension value for pure liquid mercury at 298 K in vacuum or in a variety of inert gas at 1.01325×10^5 Pa (= 1 atm) is 485 ± 1 mN/m, and its temperature coefficient appears to be -0.20 ± 0.02 mN/mK (Allen 1972*f*).
Data are taken from Lang (1973).

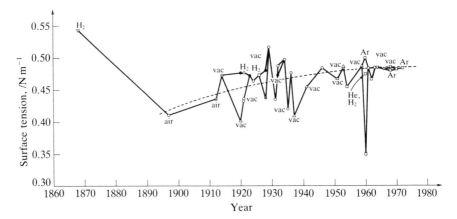

Fig. 5.2. Surface tension values for liquid mercury at room temperature as a function of year (after Morita, Iida, and Kasama 1976). Methods of surface tension measurement: ○ Sessile drop, □ Maximum bubble pressure, △ Maximum drop pressure, ● Drop weight, ▽ Pendant drop, ⊖ Oscillating jet, ⊗ Capillary depression, ① Contact angle. Data from Lang (1973).

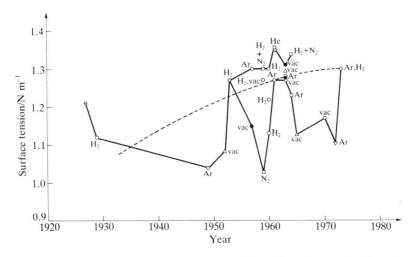

Fig. 5.3. Surface tension values for liquid copper near its melting point as a function of year. Methods of surface tension measurement: ○ Sessile drop, □ Maximum bubble pressure, △ Pendant drop, ● Drop weight, ⊖ Capillary rise. Data from Allen (1972a), and Kasama, Iida, and Morita (1976).

tension value of liquid metals shows a tendency to increase over a period of years, and to approach its presumably true value. This is supported by the fact that the extent of scatter in the values of surface tension measured gradually diminish with time. This reduced scatter can be attributed mainly to a decrease in impurities and surface contamination.

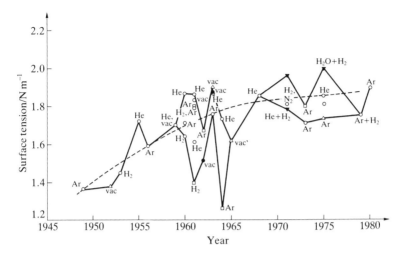

FIG. 5.4. Surface tension values for liquid iron near its melting point as a function of year. Methods of surface tension measurement: ○ Sessile drop, □ Maximum bubble pressure, △ Pendant drop, ▼ Oscillating drop, ● Drop weight. Data from Allen (1972b); Murarka, Lu, and Hamielec (1975); Kawai and Mori (1979); Ogino, Nogi, and Yamase (1980).

Considering the results of Figs. 5.2–5.4, surface tension values are largely independent of method, apart from the maximum bubble pressure technique which has frequently produced low values.

Most methods for measuring surface tensions are based on some form of the Laplace equation for describing the pressure differences set up across curved interfaces. The surface tension pressure difference ΔP is given by

$$\Delta P = \gamma \left(\frac{1}{R_1} + \frac{1}{R_2} \right) \tag{5.1}$$

where γ is the surface tension or interface tension, and R_1 and R_2 are the principal radii of curvature of the surface at the point considered. For a spherical interface, $R_1 = R_2 = R$, so that $\Delta P = 2\gamma/R$. These relations are relevant to the methods described below.

5.2.2. Methods of measurement

There are a number of methods whereby the surface tension of a liquid can be measured. For most determinations of γ, one or more of the following methods have been employed:

(a) Capillary Rise Method,
(b) Maximum Drop Method,

(c) Maximum Bubble Pressure Method,
(d) Sessile Drop Method,
(e) Pendant Drop Method,
(f) Drop Weight Method,
(g) Oscillating Drop Method.

Of these methods, the sessile drop method and the maximum bubble pressure method are most easily applied at elevated temperatures and have been most frequently used. For the high-melting-point refractory metals, the pendant drop method and the drop weight method have sometimes been employed.

The essential features of these techniques are described below.

5.2.2.1. Sessile drop method

As already mentioned in Subsection 3.2.6.1, the sessile drop method is based on the measurement of the dimensions of a stationary liquid drop (usually 3–5 $\times 10^{-7}$ m^3) resting on a horizontal substrate. This technique has been used most extensively for measuring the surface tension of liquid metals and alloys, because of the following advantages. The sessile drop method allows for more accurate determination of surface tensions over wide ranges of temperature compared with the other methods used at high temperatures. It also allows simultaneous measurements of contact angle, spreading coefficient, work of adhesion, and density. Furthermore, both the mathematical treatment for calculating surface tensions from measured drop dimensions, as well as the experimental equipment needed for producing a sessile drop, are relatively simple. However, every precaution must be taken to ensure the absence of any source of contamination. An adequate cylindrical symmetry of the sample, as well as good optical equipment,† are necessary before accurate data can be obtained.

Several equations have been presented to calculate surface tension from the dimensions of a sessile drop. Figure 5.5. shows the dimensions to be measured.

5.2.2.1.1. Bashforth and Adams equation The surface tension of the liquid specimen can be obtained using the Bashforth and Adams (1883) equation (see, e.g. Kingery and Humenik 1953):

$$\gamma = \frac{g \rho b^2}{\beta} \tag{5.2}$$

where g is the gravitational acceleration and ρ is the liquid density. From the

---

† The dimensions of a sessile drop are determined from analysis of the profile of the drop (generally, its photograph).

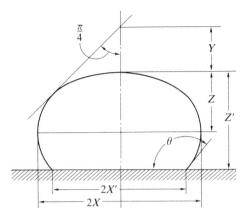

FIG. 5.5. Measurements of sessile drop for surface tension calculations.

measured values of X and Z, parameters b and β can be determined using the Bashforth and Adams (1883) tables.

The difficulty in measuring Z entails an error of 2–3 per cent on γ.

Incidentally, the drop volume V can be calculated from the relation

$$V = \frac{\pi b^2 (X')^2}{b} \left(\frac{2}{b} - \frac{2 \sin \theta}{X} + \frac{\beta Z'}{b^2} \right) \tag{5.3}$$

Density can be obtained from the drop volume and its weight.

5.2.2.1.2. Dorsey equation In this approach, the surface tension is calculated from the empirical relation, Dorsey's equation (Dorsey 1928):

$$\gamma = g \rho X^2 \left(\frac{0.0520}{f} - 0.1227 + 0.0481 \, f \right) \tag{5.4}$$

where the Dorsey factor f is given by

$$f = \frac{Y}{X} - 0.4142$$

This method yields better accuracy than the previous one, but the difficulty in the experimental measurement of Y (Fig. 5.5) remains an appreciable source of error.

5.2.2.1.3. Corrected Worthington equation

$$\gamma = \frac{1}{2} \rho g Z^2 \, \frac{1.641 \, X}{1.641 \, X + Z} \tag{5.5}$$

The mathematical treatment of this equation (Worthington 1885) appears to be satisfactory, but the difficulty in the determination of Z with high accuracy still remains.

5.2.2.2. Maximum bubble pressure method

This technique involves measuring the maximum pressure attained in bubbles formed at the tip of a capillary tube immersed to different depths within a liquid, as described in Subsection 3.2.4.

The maximum bubble pressure method has frequently been employed in experimental determinations of the surface tensions of liquid metals and alloys. Its popularity stems from the reason that each successive measurement is made on a freshly formed surface. As a result, surface contamination effects are reduced to a minimum. The method is thought to be particularly suitable for metals which are very sensitive to surface contamination, such as the alkali metals, magnesium, aluminium, and calcium, which have a great affinity for oxygen (e.g. Davies and West 1963–4). Needless to say, the inert gas, helium or argon, used to form the bubbles must be carefully purified. The measurements are carried out remotely since no direct observations on the mechanics of bubble formation in liquid metals are made. Methods have been established for obtaining absolute surface tension figures from the experimental data, which obviates any need for any experimental calibration procedure.

If the bubble detaching from the orifice tip is perfectly spherical, eqn (3.2) holds. However, since the pressure on the liquid side of the bubble varies with the head of liquid, the bubble is not spherical even though it may approach a spherical shape when a capillary tube of very small radius is used. A correction must, therefore, be applied for any distortion of the bubble due to gravitational effects. Similarly, possible non-wetting effects may lead to low values of surface tension (Irons and Guthrie 1981).

Several methods have been proposed for correcting experimental data to give absolute surface tension values. Of these, the Cantor relation corrected by Schrödinger (1915) is probably the most frequently used (Lang 1973):

$$\gamma = \frac{rP_\gamma}{2}\left\{1 - \frac{2}{3}\left(\frac{r\rho g}{P_\gamma}\right) \times 10^{-3} - \frac{1}{6}\left(\frac{r\rho g}{P_\gamma}\right)^2 \times 10^{-6}\right\} \qquad (5.6)$$

where $P_\gamma (= P_m - \rho g h)$ is in SI units, P_m is the maximum gas pressure (gauge) at an immersion depth h, and r is the capillary radius (mm).

The expression in the braces is the correction factor for P_γ. Equation (5.6) holds well for small values of $r(2\gamma/\rho g)^{-\frac{1}{2}}$, i.e. $r < 0.2(2\gamma/\rho g)^{\frac{1}{2}}$ (where $2\gamma/\rho g$ is called the 'capillary constant' or 'specific cohesion').

A method proposed by Sugden (1922, 1924), which is mathematically more complicated, holds over a wider range of capillary diameters.

This method, which makes use of two capillary tubes of different radii, has been employed on occasion. One advantage of the method is that density data are not needed in calculating surface tensions.

5.2.2.3. Pendant drop method

Figure 5.6 indicates the shape of a drop hanging from the tip of a vertical capillary or rod. If the forces of gravity and surface tension are balanced in a static pendant drop, the liquid surface tension is given (Fordham 1948; Allen 1963) by

$$\gamma = \frac{\rho g X^2}{H} \tag{5.7}$$

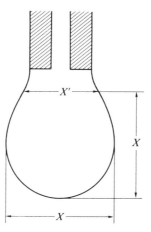

FIG. 5.6. Measurements of pendant drop for surface tension calculations (see, e.g. Allen 1963).

in which X is the maximum drop diameter and $1/H$ is the shape factor of the drop. Values of $1/H$ have been calculated as a function of the experimentally determined parameters X'/X, and usually vary between 0.3 and 1.0 (Fordham 1948). Providing equilibrium is attained and there are no errors in ρ, X', and X, eqn (5.7) yields accurate surface tension values.

In several ways, the pendant drop method is similar to the sessile drop method.

5.2.2.4. Drop weight method

When a liquid drop formed at the tip of a vertical cylindrical rod or capillary grows large enough, the liquid drop falls as a result of its weight. The liquid surface tension can be computed from the mass of the drop which falls. If m is the mass of the drop separating from a vertical rod or capillary tube of radius r as a result of gravitational forces, the surface tension of the liquid is given (Allen 1963; Harkins and Brown 1919; Calverley 1957) by

$$\gamma = \frac{mg}{2\pi r f_D} \tag{5.8}$$

where f_D is a function of $(r/V^{\frac{1}{3}})$, V being the volume of the falling drop. A drop whose weight is equal to $2\pi r\gamma$ (when $f_D = 1$, $2\pi r\gamma = mg$) is called the 'ideal drop'. Both observation and theory indicate that, while the major fraction of the drop falls, a part remains attached to the capillary.

Equation (5.8) may be rewritten as follows:

$$\gamma = \frac{mg F_D}{r} \tag{5.9}$$

in which F_D is a function of (V/r^3) or $(m/\rho r^3)$. The values of F_D corresponding to given values of V/r^3 have been listed in Tables (Harkins and Brown 1919).

The pendant drop method and the drop weight method have something in common with each other. Of the many techniques available, the pendant drop method and the drop weight method are applicable to most transition metals. In these methods, liquid drops are generally formed by electron bombardment heating in high vacuum, and their surfaces are free of any contamination. Contamination from the capillary tube can be eliminated by heating the end of a rod under study. In such procedures, since the liquid drops are suspended by solid of the same composition, these methods only permit the measurement of melting-point surface tension. The surface tension of highly reactive and refractory metals can be readily determined, provided their metals can be obtained in the form of smooth rods of a few millimeters diameter.

Neither the pendant drop technique nor the drop weight technique are suitable for determining temperature variation of γ. Furthermore, they are limited to alloys having a narrow liquid–solid range.

Evidently, great care must be taken to avoid shaking the liquid drops during the course of experiment.

5.2.2.5. Maximum drop pressure method

This method involves measuring the pressure needed to force a tiny liquid drop from the tip of an upwards-facing capillary. The technique involves similar principles to the maximum bubble pressure method.

Advantages claimed for the maximum drop pressure method are that it provides fresh and uncontaminated surfaces, avoids the introduction of contact angles, and is free of theoretical uncertainty.

This method has been employed for determining the surface tension of highly reactive, low-melting-point metals, but is limited to about 1000 K because of difficulties associated with the design of equipment for high temperatures (Hogness 1921).

5.2.2.6. Capillary rise method

This method represents an application of the well-known phenomenon of capillary action. The liquid surface tension can be computed from the expression

$$\gamma = \frac{\rho g h r}{2\cos \theta} \tag{5.10}$$

where h is the difference in height between the liquid surface within the capillary and the surface of the bulk liquid metal outside the capillary (liquid head), r is the radius of the capillary, and θ is the contact angle between the liquid and the capillary wall.

The theory of the capillary rise method is simple. However, the technique requires an exact knowledge of contact angles. Since this information is lacking for most metals, this angle is often more difficult to measure than surface tension. As a result, the method is not commonly used for metallic liquids. However, measurements of γ for some low-melting-point metals, i.e. indium, tin, and lead, have been made up to about 1000 K using this technique (e.g. Harkins and Brown 1919; Melford and Hoar 1957).

5.2.2.7. Oscillating drop method

The above-mentioned methods are static or quasi-static. A novel dynamic method known as the oscillating drop method has been used by a few investigators in the last fifteen years (see Fig. 5.4) (Fraser, Lu, Hamielec, and Murarka 1971; Murarka, Lu, and Hamielec 1971, Murarka, Lu, and Hamielec 1975). This method represents an application of the levitation technique in which the frequency of oscillation of a drop levitated in a high-frequency magnetic field (see Subsection 3.2.6.2) is measured.

It has the advantage of eliminating persistent sources of contamination which arise through the use of substrates and/or capillary tubes associated with the sessile drop, the capillary rise, and the maximum bubble pressure methods. However, it is premature to discuss quantitative aspects of the method in view of researchers' limited experience with it.

Methods for measuring surface tension of liquid metals have been described in detail in several review papers or textbooks (e.g. White 1962, 1968).

5.3. THEORETICAL EQUATIONS FOR SURFACE TENSION

Let us consider a one-component system in which two different bulk phases, a liquid and a gas phase, coexist. The presence of a gravitational field will provide a plane liquid–gas (vapour) interface. At the interface between the liquid and the gas phase, a non-uniform equilibrium phase, which is called a transition zone or an interface zone, will exist as illustrated in Fig. 5.7(a). One of the basic quantities characterizing the transition zone is the number density or the distribution function. The distribution function must vary continuously with z in the transition zone, which will exhibit a certain thickness. However, we have no means of experimentally determining either this variation of $g(r, z)$, where r denotes a vector in the (x, y) plane, or the thickness of the associated interfacial transition zone.

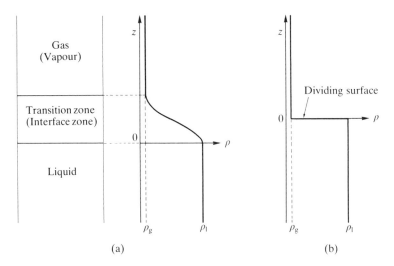

FIG. 5.7. Density variations at the interface between a liquid phase and a gas (vapour) phase: (a) liquid–gas interface region; (b) Fowler's approximation, $g(r, z) = g(r)$ for $z < 0$, $g(r, z) = 0$ for $z > 0$.

5.3.1. An equation in terms of the distribution function

A statistical mechanical analysis of the relation between surface tension and intermolecular (interatomic) forces acting at the interface between a liquid and a gas phase was first made by Fowler (1937). In the derivation of an expression for the surface tension, Fowler introduced the drastic but simple approximation that there is a mathematical surface at which a discontinuity in density exists between the liquid and the gas phase (of negligible density) as indicated Fig. 5.7(b). Fowler's expression for surface tension is given by

$$\gamma = \frac{\pi n_0^2}{8} \int\limits_0^\infty g(r) \frac{\partial \phi(r)}{\partial r} r^4 \, dr \tag{5.11}$$

Since then, Kirkwood and Buff (1949), Harashima (1953), Buff (1952), Toxvaerd (1971), and Croxton and Ferrier (1971a, 1971b) have provided further contributions to this approach. Kirkwood and Buff (1949) and Harashima (1953) have both presented rigorous expressions for surface tension in terms of the intermolecular potentials and molecular distribution functions, using a general statistical mechanical theory of interfacial phenomena. However, in their calculations of surface tension values, the approximation of a discontinuous surface between the two phases had to be introduced because of the present lack of information on the properties of the transition zone.

Johnson, Hutchinson and March (1964), and others (e.g. Waseda and Suzuki 1972) have made calculations of the surface tension values of liquid metals using experimental data available for $g(r)$ and long-range oscillatory interaction potential $\phi(r)$ through the use of Fowler's formula by eqn (5.11). Calculated values for the surface tensions of liquid metals are sometimes in quite surprisingly good agreement with experimental data. However, a close examination of the accuracy and validity of the pair interaction potentials and of Fowler's approximation is needed.

5.3.2. An equation based on fluctuation theory

Cahn and Hilliard (March and Tosi 1976), Fisk and Widom (1969), and Bhatia and March (1978a, 1978b) have developed a theory of surface tension based on (statistical) density fluctuations in non-uniform systems. According to the fluctuation theory, the surface tension can be written as the sum of two terms (Bhatia and March 1978a):

$$\gamma = \gamma_1 + \gamma_2 = \frac{l(\Delta\rho)^2}{2\rho^2 \kappa_T} + Bl\left(\frac{\Delta\rho}{l}\right)^2 \tag{5.12}$$

where l is the effective thickness of the interface, $\Delta\rho$ is the fluctuation in number density, and B is a constant. In eqn (5.12), the first term arises from treating a surface inhomogeneity as an accidental fluctuation in number density, while the second term evidently originates from the density gradient. Minimizing γ with respect to l (to determine the effective surface thickness), Bhatia and March (1978a) obtained

$$\gamma = \frac{l(\Delta\rho)^2}{\rho^2 \kappa_T} \tag{5.13}$$

or

$$\gamma \kappa_T \approx l \qquad (5.14)$$

Recently, the same authors have generalized the above approach (i.e. the Cahn and Hilliard phenomenological treatment of pure liquids) to liquid metal binary alloys Bhatia and March 1978b).

Egelstaff and Widom have presented an expression for the product of the isothermal compressibility and surface tension of a liquid near its triple point, on the basis of Fisk and Widom's theoretical studies on surface tension in the region of the critical point; (Fisk and Widom 1969; Egelstaff and Widom 1970), that is,

$$\gamma \kappa_T \approx 0.07 \, l \qquad (5.15)$$

in which L is the interface thickness. All these equations correlate the bulk property (i.e. isothermal compressibility κ_T of the liquid metal) with the surface property γ.

Equation (5.15) was verified experimentally by Egelstaff and Widom (1970) for liquids at or near their triple points. Values of $\gamma \kappa_T$ for various liquid metals are listed in Table 5.2. For liquid metals, the interface thickness $l(\approx \gamma \kappa_T/0.07)$ is in the range 2–7×10^{-10}m (2–7Å), that is, of the order of atomic size. The result demonstrates that the density gradient at the interface of a liquid metal is sharp.

An expression similar to eqns (5.14) and (5.15) can be derived from the conduction electron behaviour in metals (March and Tosi 1976).

TABLE 5.2. *Values of $\gamma \kappa_T$ for various liquid metals at or near their triple points*

Metal	γ $(10^{-3} \, \text{Nm}^{-1})$	κ_T $(10^{-11} \, \text{m}^2 \, \text{N}^{-1})$	$\gamma \kappa_T$ $(10^{-10} \, \text{m})$
Alkali metals			
Sodium	194	21	0.40
Potassium	113	40	0.45
Rubidium	95	49	0.46
Caesium	71	67	0.47
Other metals			
Iron	1790	1.43	0.25
Copper	1280	1.45	0.19
Silver	940	1.86	0.18
Zinc	785	2.4	0.19
Cadmium	666	3.2	0.21
Lead	470	3.5	0.17
Bismuth	395	4.3	0.17

Data from Egelstaff and Widom (1970).

5.3.3. An equation based on a hard-sphere model

An expression has been derived for the surface tension of a hard-sphere fluid which takes the form (March and Tosi 1976; Reiss, Frisch, and Lebowitz 1959)

$$\gamma = \frac{9kT\eta^2(1+\eta)}{2\pi\sigma^2(1-\eta)^3} \qquad (5.16)$$

where σ is an effective hard-sphere diameter and η is the packing fraction.

A correlation between surface tension and isothermal compressibility yields (Mayer 1963; Egelstaff and Widom 1970)

$$\gamma\kappa_T \approx \frac{\sigma(2-3\eta+\eta^3)}{4(1+2\eta)^2} \qquad (5.17)$$

As already stated, the packing fraction is approximately equal to 0.45 for many liquid metals, so that eqn (5.17) yields

$$\gamma\kappa_T \approx 0.05\sigma \qquad (5.18)$$

5.3.4. An equation based on a free-electron model

Several workers (Frenkel 1917; Gogate and Kothari 1935; Brager and Schuchowitzky 1946; Huang and Wyllie 1949; Stratton 1953; Taylor 1955) have developed a theory of surface tension or surface energy based on a free-electron gas model of a liquid metal. Gogate and Kothari (1935) considered the surface layer of liquid metals to be a two-dimensional electron gas which obeyed Fermi–Dirac statistics. Their resulting expression for the surface tension of liquid metals is that γ is inversely proportional to the liquid metal's atomic volume according to

$$\gamma \propto V^{-\frac{4}{3}} \qquad (5.19)$$

Figure 5.8 indicates that this correlation holds roughly for a large number of metallic liquids.

Stratton (1953) has since refined the theoretical calculation of the surface energy of liquid metals by taking into account the phenomenon of charge conservation at the metal surface. Calculated surface energy values for the alkali metals by Stratton appear to reproduce experimental data with good agreement (Taylor 1954–5, 1955). However, for other metals, the agreement between calculation and experiment is generally poor, owing to the simplicity of the free-electron model when applied to the more complex electron structures of the transition metals, etc.

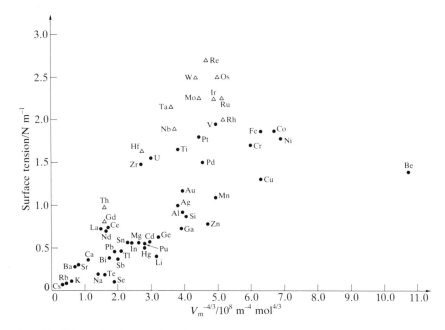

FIG. 5.8. Values of surface tension of various liquid metals at their melting points vs. (atomic volume)$^{-\frac{4}{3}}$ $\triangle$ Estimated density values (Allen 1972c) are used.

5.4. SEMI-EMPIRICAL AND EMPIRICAL EQUATIONS FOR SURFACE TENSION

(1) Atterton and Hoar (1951) have demonstrated (for 18 metals) that surface tension is approximately inversely proportional to atomic volume, and that the relationship holds for metals of such widely different types as sodium, bismuth, and iron. As already stated, the relationship between those two properties is by no means unexpected on theoretical grounds. However, Atterton and Hoar's empirical relationship exhibits a rather large scatter for many metallic liquids, as can be seen from Fig. 5.8.

(2) Skapski (1948) and, shortly after, Oriani (1950) have proposed expressions for the surface tension of liquids in terms of their enthalpies of sublimation using the quasi-chemical method.

According to their treatment of surface tension, the total molar surface energy of a liquid is determined by the amount of energy required to bring N_A atoms (Avogadro's number of atoms) to its free surface. The total molar surface energy γ_0 of the liquid can be represented as

$$\gamma_0 = S_A\left\{\gamma - T\left(\frac{d\gamma}{dT}\right)\right\} \qquad (5.20)$$

in which S_A is the surface area occupied by a monatomic layer of N_A atoms. Such a surface S_A is given by

$$S_A = fN_A\left(\frac{V}{N_A}\right)^{\frac{2}{3}} = fN_A^{\frac{1}{3}}\left(\frac{M}{\rho}\right)^{\frac{2}{3}} \tag{5.21}$$

where f is a surface-packing or configuration factor.[†] This energy, i.e. the total molar surface energy, can be calculated directly from the enthalpy of sublimation at $0\,\mathrm{K}, \Delta_s^g H_0$, and from the configuration of nearest-neighbour atoms (the quasi-chemical method).

Oriani's expression for the surface tension or the total molar surface energy is given (Oriani 1950) by

$$\gamma = \frac{1}{S_A}\left\{\left(\frac{Z_i - Z_s}{Z_i}\right)\Delta_s^g H_0 - \frac{Z_s\phi}{2}\right\} + T\left(\frac{\mathrm{d}\gamma}{\mathrm{d}T}\right) \tag{5.22}$$

or

$$\gamma_0 = \left(\frac{Z_i - Z_s}{Z_i}\right)\Delta_s^g H_0 - \frac{Z_s\phi}{2} \tag{5.23}$$

in which Z_i is the coordination number within the interior or bulk of the liquid, Z_s is the equivalent coordination number at the surface of the liquid, and ϕ, the excess binding energy, is defined as the difference in the pairwise interaction energy among atoms within the bulk of the liquid, u, and that on the surface layer, v, (i.e. $\phi \equiv v - u$).

By assuming that the pairwise binding energy of surface atoms is equal to that for atoms within the interior of the liquid, i.e. $\phi = v - u = 0$, eqns (5.22) and (5.23) become identical to Skapski's expressions for γ and γ_0.

A plot of the molar surface energy γ_0 versus the enthalpy of sublimation $\Delta_s^g H_0$ of various metallic liquids, extrapolated to absolute zero, is presented by Allen (1972e). The correlation between γ_0 and $\Delta_s^g H_0$ exhibits a similar scatter to that of Figs. 5.11 or 5.13 (a spread of ± 20 per cent or more).

(3) By assuming a simple function for the interatomic potential, Iida, Kasama, Misawa, and Morita (1974) have derived an expression for the surface tension of liquid metals in which they consider the work required to separate atoms to a distance at which interatomic forces become negligible.

We first present an expression for the surface tension of liquid metals at their melting points. To simplify the treatment, let us consider a homogeneous

---

[†] For close-packed liquids, $f = 1.09$; for b.c.c. liquids (coordination number 8), $f = 1.12$; for liquid mercury (coordination number 6), $f = 1.04$.

liquid metal at its melting point, which consists of atoms of mass m, performing harmonic vibrations of the same frequency v_1 (Einstein frequency) about positions of equilibrium.

The (Einstein) frequency is given by

$$v_1 = \frac{1}{2\pi}\left(\frac{K_f}{m}\right)^{\frac{1}{2}} \tag{5.24}$$

where K_f is the force constant. In order to represent the force constant in terms of well-known physical parameters, we employ a modified version of Lindemann's melting formula for atomic frequency, this being

$$v_1 = \beta v_L = \beta c \left(\frac{RT_m}{M V_m^{\frac{2}{3}}}\right)^{\frac{1}{2}} \qquad (c = 3.0 \times 10^8) \tag{5.25}$$

in which β is a correction factor. From eqns (5.24) and (5.25), we have for the force constant,

$$K_f = 4\pi^2\beta^2 c^2 \frac{RT_m}{N_A V_m^{\frac{2}{3}}} \tag{5.26}$$

The interatomic force $f(s)$ acting between a pair atoms can be expressed by

$$f(s) = -ks$$

where

$$k = K_f/Z_i \tag{5.27}$$

In eqn (5.27), Z_i is the nearest-neighbour coordination number of atoms in the bulk of the liquid [†], and s is the distance of displacement of the central atom from its position of equilibrium.

The next step is to select an arbitrary dividing surface in the interior of the liquid metal and to define a rectangular coordinate system (x, y, z) in which the dividing surface lies in the (x, y) plane, as depicted in Fig. 5.9. Consider an atom 1, situated at a point $z = 0$ on the dividing surface, i.e. the (x, y) plane, and an atom 2, situated below the (x, y) plane at the equilibrium separation distance, or the average interatomic distance, a_m from atom 1. When we displace atom 1 in the direction of the z axis by a distance z, the component of force f_z in the z direction acting between the two atoms separated by a distance $r_{1 \cdot 2}$ is given by

$$f_z = -k(\sqrt{z^2 + 2a_m z \cos\theta + a_m^2} - a_m)\cos\phi \tag{5.28}$$

[†] $\dfrac{1}{2}Z = \displaystyle\int_0^{r_1}\int_0^{\pi/2} 2\pi r^2 n_0 g(r) \sin\theta \, d\theta \, dr$, where r_1 is the maximum value of nearest-neighbour distance.

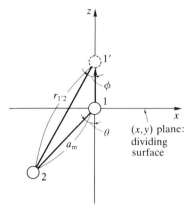

FIG. 5.9. A model for the explanation of surface tension.

where

$$\cos\phi = \frac{z + a_{\mathrm{m}}\cos\theta}{\sqrt{z^2 + 2a_{\mathrm{m}}z\cos\theta + a_{\mathrm{m}}^2}}$$

and subscript m refers to the melting point.

For simplicity, we now introduce the drastic assumption that the potential of the interatomic force acting between a pair of atoms separated by a distance r may be expressed in the form illustrated in Fig. 5.10. On the assumption that

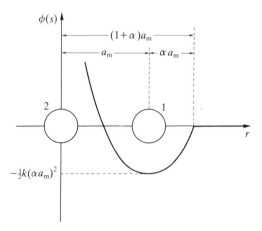

FIG. 5.10. A schematic diagram of the pair potential for deriving an expression for the surface tension of liquid metals at their melting points.

this potential energy function applies, the distance z_1 over which the interatomic force extends is

$$\sqrt{z_1^2 + 2a_m z_1 \cos\theta + a_m^2} = (1 + \alpha)a_m \qquad (5.29)$$

or

$$z_1 = a_m[\{\cos^2\theta + \alpha(\alpha + 2)\}^{\frac{1}{2}} - \cos\theta] \qquad (5.30)$$

Consequently, the work or the energy w_m necessary to separate atom 1 to a distance beyond which the interatomic forces between the two atoms (1 and 2) are no longer felt (i.e. the force declines to zero) can be expressed as follows

$$w_m = -\int_0^{z_1} f_z\,dz = \int_0^{z_1} k\{(z^2 + 2a_m z \cos\theta + a_m^2)^{\frac{1}{2}} - a_m\}$$

$$\times \frac{z + a_m \cos\theta}{(z^2 + 2a_m z \cos\theta + a_m^2)^{\frac{1}{2}}}\,dz = \frac{1}{2}k(\alpha a_m)^2 \qquad (5.31)$$

Combining eqns (5.26), (5.27), and (5.31), we have

$$w_m = \frac{2(\pi\alpha\beta c a_m)^2\, R T_m}{N_A Z_i V_m^{\frac{2}{3}}} \qquad (5.32)$$

If we neglect the work needed to change the distribution of atoms in the bulk of the liquid into their new distribution within the surface transition zone (the same approximation was introduced by Fowler (1937)), we have for the surface tension of liquids at their melting points, γ_m

$$\gamma_m \approx \frac{Z_i w_m}{4a_m^2} \qquad (5.33)$$

In eqn (5.33), $(1/a_m^2)$ and $(Z_i/2)$ respectively represent the average number of atoms per unit area on the dividing surface, and the number of equivalent atoms to atom 2 in the correlation between the atoms 1 and 2. By substituting eqn (5.32) into (5.33), we have for the surface tension of liquid metals at their melting points,

$$\gamma_m \approx \frac{(\pi\alpha\beta c)^2}{2N_A} \frac{R T_m}{V_m^{\frac{2}{3}}} = 7.9 \times 10^{-7}(\alpha\beta)^2 \frac{R T_m}{V_m^{\frac{2}{3}}} \qquad (5.34)$$

Similar relationships to eqn (5.34) have already been proposed by Schytil (1949) using a different theoretical viewpoint, and by Allen (1972f) using a semi-empirical treatment. Both authors have indicated that the surface tension of liquid metals at their melting points is approximately proportional to (melting temperature T_m)/(atomic volume V_m)$^{\frac{2}{3}}$, as shown in Fig. 5.11. This

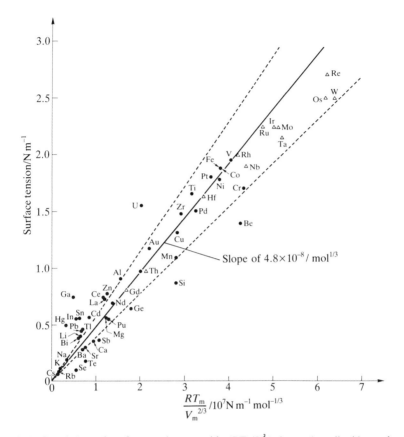

FIG. 5.11. Correlation of surface tension γ_m with $(RT_m/V_m^{\frac{2}{3}})$ for various liquid metals. $\triangle$ Estimated density values (Allen 1972c) are used.

implies that the product $\alpha\beta$ must be nearly equal for all metallic elements. From the relationship indicated Fig. 5.11, we obtain

$$\gamma_m \approx 4.8 \times 10^{-8}\, \frac{RT_m}{V_m^{\frac{2}{3}}}, \quad \text{(in SI units)} \tag{5.35}$$

and

$$\alpha\beta = 0.25 \tag{5.36}$$

Second, if we assume that, at any temperature above the melting point, the isothermal work w needed to create unit area of free surface is reduced by the work of thermal expansion (Kasama, Iida, and Morita 1976), w can be expressed by the relation (Fig. 5.12)

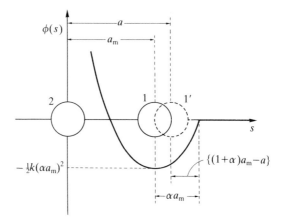

FIG. 5.12. A schematic diagram of the pair potential for the explanation of temperature variations of surface tension.

$$w = \frac{1}{2} k\{(1 + \alpha)a_m - a\}^2 \tag{5.37}$$

Following equivalent procedures to those described above for the surface tension of liquid metals at their melting points, the surface tension of liquid metals at any temperature above their melting points can then be expressed by

$$\gamma \approx \frac{(\pi\beta c)^2 \, T_m}{2N_A} \frac{\{(1 + \alpha) V_m^{\frac{1}{3}} - V^{\frac{1}{3}}\}^2}{V^{\frac{4}{3}}} \tag{5.38}$$

or

$$\gamma \approx \frac{(\pi\beta c)^2 \, T_m}{2N_A M^{\frac{2}{3}}} \left(\frac{\rho}{\rho_m}\right)^{\frac{2}{3}} \{(1 + \alpha)\rho^{\frac{1}{3}} - \rho_m^{\frac{1}{3}}\}^2 \tag{5.39}$$

If eqns (5.38) and (5.39) hold over the entire range of liquid metal temperatures, we can obtain a value for the parameter α since surface tension values reduce to zero at the critical point

$$\alpha = \left(\frac{\rho_m}{\rho_c}\right)^{\frac{1}{3}} - 1 = \left(\frac{V_c}{V_m}\right)^{\frac{1}{3}} - 1 \tag{5.40}$$

As stated in Subsection 4.4.4, the value of α is approximately equal to 0.51 for liquid metals. Consequently, the correction factor β for Lindemann's melting formula can be evaluated from the expression:

$$\beta = \frac{1.1 \times 10^3 \, V_m^{\frac{1}{3}}}{\alpha} \left(\frac{\gamma_m}{R T_m}\right)^{\frac{1}{2}} = 2.2 \times 10^3 \, V_m^{\frac{1}{3}} \left(\frac{\gamma_m}{R T_m}\right)^{\frac{1}{2}} \tag{5.41}$$

From eqn (5.25), the mean atomic frequency of liquid metals at their melting points can therefore be expressed as

$$\nu_1 = \beta \nu_L = 6.8 \times 10^{11} \left(\frac{\gamma_m}{M}\right)^{\frac{1}{2}} \tag{5.42}$$

As mentioned previously, the velocity of sound in liquid metals has been successfully represented through the use of the parameter β. The relationship by eqn (4.44) can be rewritten in terms of the isentropic compressibility κ_S and the isothermal compressibility κ_T (eqn (4.30)) as follows:

$$\gamma_m \kappa_{S,m} \approx 6.9 \times 10^{-10} V_m^{\frac{1}{3}} \tag{5.43}$$

$$\gamma_m \kappa_{T,m} \approx 6.9 \times 10^{-10} \left(\frac{C_p}{C_V}\right)_m V_m^{\frac{1}{3}}. \tag{5.44}$$

where the subscript m represents the melting point. On combining eqn (5.44) and the relation of packing fraction for liquid metals at their melting points, i.e. $\eta_m = \pi N_A \sigma^3 / 6 V_m \approx 0.45$, we have

$$\gamma_m \kappa_{T,m} \approx 0.07\sigma \tag{5.45}^{\dagger}$$

These equations are similar to independently derived relations by eqns (5.14), (5.15), and (5.18).

(4) From the relationship given in Fig. 1.5, or Skapski's equation (see subsection 5.4, (2)), a simple correlation between surface tension γ_m and enthalpy of evaporation $\Delta_l^g H_m$ can be expected. Figure 5.13 indicates that the correlation of γ_m with $(\Delta_l^g H_m / V_m^{\frac{2}{3}})$ is approximately true for a large number of metals. This correlation is expressed by

$$\gamma_m \approx 1.8 \times 10^{-9} \frac{\Delta_l^g H_m}{V_m^{\frac{2}{3}}} \tag{5.46}$$

5.5. TEMPERATURE COEFFICIENTS OF SURFACE TENSION

Since the surface between liquid phase and gas phase disappears at the critical temperature T_c, it follows that the surface tension of liquids is reduced to zero at T_c. As a consequence, the surface tension of liquids must decrease with rising temperature.

A well known relationship between surface tension and absolute temperature is Eötvös' law, which can be expressed in the form

$$\gamma = \frac{k_\gamma}{V^{\frac{2}{3}}} (T_c - T) \tag{5.47}$$

---

† According to Kleppa's experimental investigations, γ_h ($\equiv C_p/C_V$) is approximately equal to 1.15 for all liquid metals (Kleppa 1950).

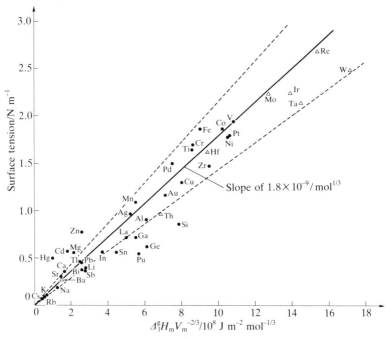

FIG. 5.13. Correlation of γ_m with $(\Delta_l^g H_m/V_m^{\frac{2}{3}})$ for various liquid metals. $\triangle$ Estimated density values (Allen 1972c) are used.

where k_γ is approximately equal to 6.4×10^{-8} $(\mathrm{J\,K^{-1}\,mol^{-\frac{2}{3}}})$ for all liquid metals. By differentiating eqn (5.47) with respect to temperature T, we have for the temperature coefficient of surface tension, $\mathrm{d}\gamma/\mathrm{d}T$, that

$$\frac{\mathrm{d}\gamma}{\mathrm{d}T} = \frac{k_\gamma}{V^{\frac{2}{3}}}\left\{\frac{2(T_c - T)}{3\rho}\frac{\mathrm{d}\rho}{\mathrm{d}T} - 1\right\} = \frac{\gamma}{T_c - T}\left\{\frac{2(T_c - T)}{3\rho}\frac{\mathrm{d}\rho}{\mathrm{d}T} - 1\right\} \quad (5.48)$$

Using this relationship, Allen (1972c, 1972f) has calculated values of $\mathrm{d}\gamma/\mathrm{d}T$ for high-melting-point metals for which no experimental data exist, in which estimated values of T_c by Grosse are employed (see Subsection 3.3.1).

On substituting the density–temperature correlation, i.e. $\rho = \rho_m + \Lambda T$, into eqn (5.39) and differentiating γ with respect to T, we have

$$\frac{\mathrm{d}\gamma}{\mathrm{d}T} = \frac{(\pi\beta c)^2 \Lambda T_m}{3 N_A M^{\frac{2}{3}}}\left\{2(1+\alpha)^2 \rho^{\frac{1}{3}}\rho_m^{-\frac{2}{3}} + \rho^{-\frac{1}{3}} - 3(1+\alpha)\rho_m^{-\frac{1}{3}}\right\} \quad (5.49)$$

In Table 5.3, calculated $(\mathrm{d}\gamma/\mathrm{d}T)$ values from eqn (5.49) are compared with experimental data (Kasama, Iida, and Morita 1976). Agreement with experimental data is seen to be good.

TABLE 5.3. *Comparison of the calculated and observed temperature coefficients of surface tension for various liquid metals*

Metal	$(d\gamma/dT)_{obs}$ $(mN\,m^{-1}\,K^{-1})$	$(d\gamma/dT)_{calc}$ $(mN\,m^{-1}\,K^{-1})$
Li	-0.14	-0.16
Na	$-0.05\_-0.10$	-0.14
K	$-0.06\_-0.11$	-0.07
Rb	-0.06	-0.06
Cs	-0.06	-0.05
Cu	$+0.68\_-0.31$	-0.28
Ag	$-0.13\_-0.19$	-0.24
Au	$-0.10\_-0.52$	-0.27
Mg	$-0.30\_-0.35$	-0.24
Ca	-0.10	-0.14
Ba	-0.08	-0.06
Zn	$-0.17\_-0.25$	-0.25
Cd	-0.26 (not linear)	-0.23
Hg	-0.20	-0.22
Al	$-0.14\_-0.35$	-0.26
Ga	-0.10	-0.18
In	-0.09	-0.17
Tl	-0.08	-0.13
Ge	-0.26	-0.14
Sn	$-0.02\_-0.22$	-0.14
Pb	$-0.06\_-0.24$	-0.13
Sb	$0.00\_-0.05$	-0.09
Bi	-0.13	-0.11
Fe	$-0.02\_-0.50$	-0.47
Co	-0.38	-0.40
Ni	-0.49	-0.45

Data from Kasama *et al.* (1976); Morita, Iida, and Kasama (1976).

One should note that temperature coefficients of surface tension calculated through the use of a purely theoretical expression, i.e. Fowler's expression, coincide qualitatively with experimental values.

5.6. EXPERIMENTAL DATA FOR PURE LIQUID METALS

Table 5.4 lists experimental values of surface tension for pure liquid metals at their melting points, together with their temperature coefficients of surface tension. The values shown have been extrapolated to their melting points on the basis of experimental measurements made at elevated temperatures. The

TABLE 5.4. *Surface tensions of pure liquid metals at their melting points*

Metal	γ (mN m^{-1})	$d\gamma/dT$ (mN m^{-1} K^{-1})	Metal	γ (mN m^{-1})	$d\gamma/dT$ (mN m^{-1} K^{-1})
Li	398	−0.14	Pd	1500	(−0.22) {−0.45}
Be	1390	(−0.29)	Ag	966[b]	−0.19[b]
B	1060[a]	—	Cd	570	−0.26 (not linear)
Na	191	−0.10	In	556	−0.09
Mg	559	−0.35	Sn	560[b]	−0.09[b]
Al	914	−0.35	Sb	367	−0.05
Si	865	(−0.13)	Te	180	(−0.06)
K	115	−0.08	Cs	70	−0.06
Ca	361	−0.10	Ba	277	−0.08
Ti	1650	(−0.26) {−0.24}	La	720	−0.32
V	1950	(−0.31) {−0.30}	Ce	740	−0.33
Cr	1700	(−0.32)	Nd	689	−0.09
Mn	1090	−0.2	Gd	810	(−0.16)
Fe	1872	−0.49	Hf	1630	(−0.21) {−0.22}
Co	1873	−0.49	Ta	2150	(−0.25) {−0.26}
Ni	1778	−0.38	W	2500	(−0.29) {−0.30}
Cu	1303[b]	−0.23[b]	Re	2700	(−0.34) {−0.31}
Zn	782	−0.17	Os	2500	(−0.33) {−0.64}
Ga	718	−0.10	Ir	2250	(−0.31) {−0.58}
Ge	621	−0.26	Pt	1800	(−0.17) {−0.70}
Se	106	−0.1	Au	1169[b]	−0.25[b]
Rb	85	−0.06	Hg	498	−0.20
Sr	303	−0.10	Tl	464	−0.08
Zr	1480	(−0.20) {−0.21}	Pb	458[b]	−0.13[b]
Nb	1900	(−0.24) {−0.25}	Bi	378	−0.07
Mo	2250	(−0.30) {−0.31}	Th	978	(−0.14)
Ru	2250	(−0.31) {−0.58}	U	1550	−0.14
Rh	2000	(−0.30) {−0.52}	Pu	550	(−0.10)

[a] Data from Wilson (1965c).
[b] Data from Kasama *et al.* (1976).
Values in parentheses and in braces were calculated from eqns (5.48) (Allen 1972c) and (5.49) (Morita, Iida, and Kasama 1976).
All other data from Allen (1972c).

data sources for these values come from the works of Grosse (1964), Wilson (1965c), Allen (1972c) and Kasama, Iida, and Morita (1976).

A discussion of such values is available in the review paper of Allen (1972f) and will not repeated here. However, we wish to emphasize the effects of impurities on the surface tension values of liquid metals. Most solute elements, particularly non-metals such as oxygen and sulphur, are highly surface active in liquid metals. The effects of these dissolved elements on the surface tensions of liquid metals can be remarkably large. As a result, the reported values for the surface tension of many liquid metals must be regarded as being of poor reliability, unless special precautions were taken to eliminate any source of surface contamination.

It is extremely difficult to assign an overall error to surface tension measurements, but $\pm 5 - \pm 10$ per cent would seem to be a fair estimate, considering the review paper of Allen (1972f) and the results shown in Figs. 5.2–5.4.

Concerning the values of $d\gamma/dT$ for pure liquid metals, experimental results confirm the negative coefficients expected theoretically. Data for the values of $d\gamma/dT$ listed in Table 5.3 can generally be expressed in terms of linear relationship with respect to temperature, with negative coefficients lying in the order of $-0.1 - -0.7 \times 10^{-3} \, \mathrm{N\,m^{-1}\,K^{-1}}$. The accuracies of the temperature coefficients therein listed are estimated to be about ± 50 per cent.

For cadmium, zinc, and copper, there are sufficient data to indicate positive temperature coefficients over limited temperature ranges (Allen 1972f; White 1966, 1972; Mittag and Lange[†] 1975). Most of the positive coefficients reported can be attributed to impurity effects or measurements made under non-equilibrium conditions. However, White (1972) insists that temperature coefficients are positive when the degree of atomic order in the surface is higher than in the bulk of a liquid. This is the case with cadmium.

5.7. ADSORPTION OF SOLUTES ON LIQUID METAL SURFACES

The adsorption behaviour of solute in liquid metals has, and can be, investigated by measuring the surface tension of a liquid metal as a function of solute concentration. The results can then be interpreted on the basis of a thermodynamic treatment of interfaces.

Thus, the excess surface concentration of solute in a two-component (binary) system at constant temperature and pressure is given (e.g. Moelwyn-Hughes 1961a) by

$$\Gamma_s = -\frac{d\gamma}{RT\,d\,(\ln a_s)} \tag{5.50}$$

in which Γ_s is the excess quantity of solute s associated with unit area of surface, i.e. the excess surface concentration per unit area and a_s is the activity of solute s in the system. Equation (5.50) is known as the Gibbs adsorption equation or the Gibbs adsorption isotherm. In dilute solutions, where Henry's law is obeyed, the solute's activity a_s can be replaced by the solute's concentration in terms of weight per cent or atom per cent. In other words, at low concentrations of solute, Γ_s can be taken to equal the surface concentration of solute per unit interfacial area. As is evident from eqn (5.50), the excess surface concentration Γ_s can be evaluated from the slope of experimentally determined $d\gamma/d\,(\ln a_s)$ for $d\gamma/d\,(\ln x)$ values, where x is atomic fraction.

[†] The surface tension (the interface tension between pure iron and an argon atmosphere) of iron increases with increasing temperature.

Halden and Kingery (1955) have carried out experimental determinations of the effect of carbon, nitrogen, oxygen, and sulphur additions on the surface tension of liquid iron. These surface tension measurements are plotted as a function of ln (wt.%) in Fig. 5.14. At these concentrations, the activity is essentially equal to concentration for all materials apart from carbon.

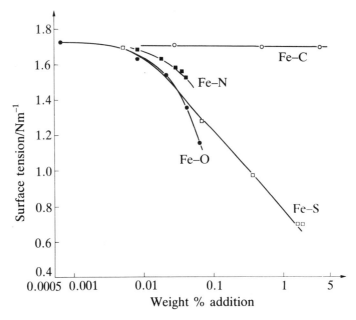

FIG. 5.14. Effect of C, N, S and O on the surface tension of liquid iron (after Halden and Kingery 1955).

Halden and Kingery (1955) have deduced excess surface concentrations of solutes using the slopes of the curves shown in Fig. 5.14. Their calculated values are shown in Fig. 5.15. As is obvious from this figure, they obtained the following results. The excess surface concentration of oxygen rapidly increases up to a value of 21.8×10^{-6} mol m^{-2} at about 0.04 per cent. The area per oxygen atom at the surface is 7.62×10^{-20} m^2 (7.62 Å^2), in reasonable agreement with the value of 8.12×10^{-20} m^2 per atom found in FeO for the plane of maximum packing, and with the value of 6.78×10^{-20} m^2 calculated from Pauling's radius of 1.40×10^{-10} m for O$^=$.

Referring now to sulphur dissolved in iron, its excess surface concentration also rapidly increases up to a value of 11.6×10^{-6} mol m^{-2}. This concentration corresponds to an area of 14.4×10^{-20} m^2 per atom, which is somewhat larger than the value of 11.56×10^{-20} m^2 per atom in the plane of maximum

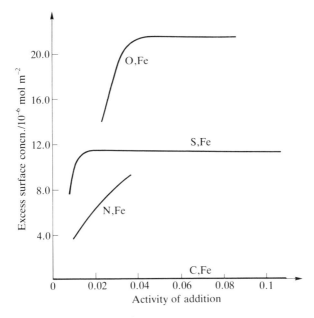

F IG. 5.15. Excess surface concentration of additions to iron at 1843 K (after Halden and Kingery 1955).

packing for FeS as well as the value of $10.49 \times 10^{-20} \, m^2$ calculated from Pauling's radius of $1.84 \times 10^{-10} \, m$ for $S^=$. At low concentrations, sulphur is seen to be more highly surface active than oxygen. This is due to its large ionic size, which leads to the ion becoming highly polarized by the iron's ionic potential. Similarly, complete surface coverage occurs more rapidly with sulphur, and at a lower activity, than with oxygen.

Nitrogen has a smaller effect on surface tension, decreasing the surface tension of iron by only $0.2 \, N \, m^{-1}$ at $1.01 \times 10^5 \, Pa$ (1 atm) of nitrogen. At this concentration, the excess surface concentration was $8.3 \times 10^{-6} \, mol \, m^{-2}$, which is only a small fraction of a hexagonal close-packed monolayer $(95 \times 10^{-6} \, mol \, m^{-2})$. Carbon has no effect on the surface tension of pure iron at 1843 K.

Subsequent to the work of Halden and Kingery, several investigators made experimental determinations of the effects of controlled solute additions on the surface tension of liquid iron. Part of their results is summarized in Fig. 5.16 by Allen (1972d), in which surface tension is plotted versus solute concentration, in weight per cent. As seen, there are no large discrepancies among the data. However, measured values for the surface tension of pure liquid iron would appear to be slightly low on the basis of simple extrapolation of the curves

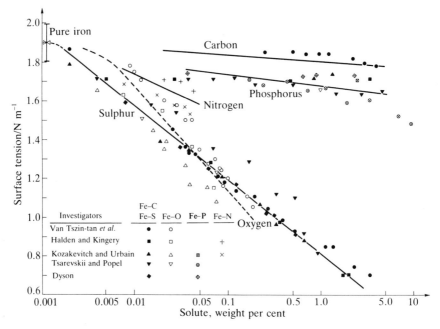

FIG. 5.16. Effect of various non-metals on the surface tension of liquid iron at 1823–1843 K (after Allen 1972*d*).

and/or lines in Fig. 5.16, even though careful measurements were undoubtedly made.

Other examples for the effects of non-metallic and metallic additions on the surface tension of liquid iron are given in Figs. 5.17 and 5.18.

Olsen and Johnson (1963) have studied the surface tensions of mercury–thallium amalgams as a function of thallium content. Their experimental results are shown in Figs. 5.19 and 5.20. They determined the excess surface concentration Γ_{Tl} from the slope of the plot of the surface tension value for the amalgams, versus the logarithm of the mole fraction of thallium in mercury, i.e. ln x_{Tl} (Fig. 5.20). As seen in Fig. 5.20, a linear relation exists for this system at mole fractions below 0.084 (because of similar atomic weights, the mole fraction values of mercury–thallium solutions are approximately equal to their weight fractions). Analysis of the graph for this region yielded an excess surface concentration Γ_{Tl} of 1.78×10^{-6} mol m^{-2}. Olsen and Johnson's explanation for these results is as follows. The limiting concentration expected in a close-packed monolayer of thallium atoms with an atomic radius of 1.99 $\times 10^{-10}$ m is estimated to be 12×10^{-6} mol m^{-2}. A comparison of these values suggests that either an imperfect monolayer is formed or that the

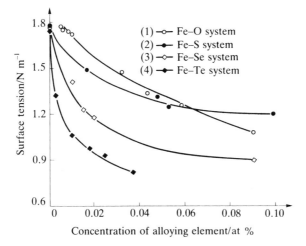

F IG. 5.17. Effect of alloying elements on the surface tension of liquid iron at 1873 K (after Ogino, Nogi, and Yamase 1980).

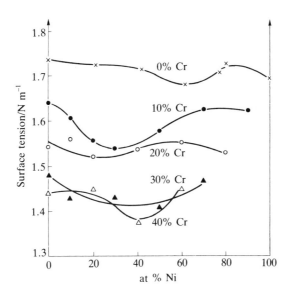

F IG. 5.18. Surface tension of iron–nickel–chromium alloys at 1823 K (after Mori, Kishimoto, Shimose, and Kawai 1975).

assumption of close packing in the monolayer is incorrect. Thus, at concentrations less than 8.5 wt. %, the thallium appears to concentrate in a surface layer on the mercury with an accompanying reduction in the surface tension of the amalgam. The increase in surface tension for amalgams with thallium content

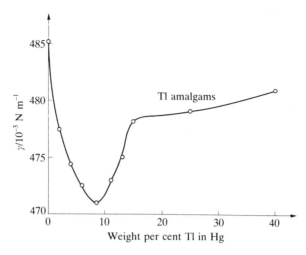

FIG. 5.19. Surface tension vs. weight per cent thallium in mercury (after Olsen and Johnson 1963).

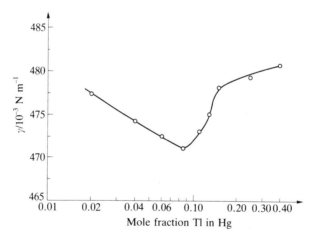

FIG. 5.20. Surface tension vs. logarithm of mole fraction thallium in mercury (after Olsen and Johnson 1963).

greater than that of the eutectic composition is more difficult to explain[†]. However, if there are components from compounds which are less stable in the surface layer than in the bulk, the surface tension of the mixture may be higher

[†] For a number of systems, the slope remains constant over a considerable composition range, corresponding to adsorption of a monolayer at the surface (Halden and Kingery 1955; Olsen and Johnson 1963).

than that of the pure components. It would appear that a compound $(Hg_x Tl_y)$ is formed which might be concentrated in the bulk of the amalgam. The formation of such a compound would remove thallium atoms from the surface layers and thereby raise surface tension values.

Thus, surface tension data for a solution as a function of concentration can offer potentially useful information. For example, in the rate of absorption of nitrogen into liquid iron, Belton (1976) considered the adsorption of strongly surface active solutes in terms of the ideal Langmuir isotherm. The Langmuir isotherm[†] is most simply expressed, at constant temperature, and for a single solute forming a monolayer as follows:

$$\frac{\theta_s}{1 - \theta_s} = K\, a_s \qquad (5.51)$$

Here θ_s is the fractional coverage by the single solute and K is called a coverage independent adsorption coefficient. Belton (1976) has shown that combination of the Langmuir and Gibbs isotherms, assuming ideal isotherms hold at all compositions, leads to (see March and Tosi 1976).

$$\gamma_0 - \gamma = RT\Gamma_s^0 \ln\left(1 + K\, a_s\right) \qquad (5.52)$$

in which γ_0 is the surface tension of the pure solvent, $(\gamma_0 - \gamma)$ is the depression of surface tension of the pure liquid metal, and Γ_s^0 is the saturation coverage by the solute $(\theta_s = \Gamma_s/\Gamma_s^0)$. Belton (1976) has reported that eqn (5.52) can be used in conjunction with experimental values of Γ_s^0 and K to give a very good description of the experimental results for surface-tension–concentration curves, and has presented the following isotherms (in units of mN/m).

Fe–S solutions at 1823 K

$$1788 - \gamma = 195 \ln\left(1 + 185 a_S\right) \qquad (5.53)$$

Fe–C(2.2 wt. %)–S solutions at 1823 K

$$1765 - \gamma = 184 \ln\left(1 + 325 a_S\right) \qquad (5.54)$$

Fe–O solutions at 1823 K

$$1788 - \gamma = 240 \ln\left(1 + 220 a_O\right) \qquad (5.55)$$

Fe–Se solutions at 1823 K

$$1788 - \gamma = 176 \ln\left(1 + 1200 a_{Se}\right) \qquad (5.56)$$

[†] The Langmuir isotherm arises from the assumption that the energy of adsorption of the species is independent of surface coverage and atomic arrangement of the adsorbing species on the surface. Incidentally, the same equation can be derived through statistical mechanics (e.g. Moelwyn-Hughes 1961a).

Cu–S solutions at 1393 K

$$1276 - \gamma = 132 \ln (1 + 140a_S) \qquad (5.57)$$

Cu–S solutions at 1473 K

$$1263 - \gamma = 140 \ln (1 + 64a_S) \qquad (5.58)$$

Cu–S solutions at 1573 K

$$1247 - \gamma = 149 (1 + 27a_S) \qquad (5.59)$$

Ag–O solutions at 1253 K

$$923 - \gamma = 50 \ln (1 + 340a_O) \qquad (5.60)$$

Ag–O solutions at 1380 K

$$904 - \gamma = 55 \ln (1 + 57a_O) \qquad (5.61)$$

As examples, the values calculated by the above equations are shown together with experimental data in Figs. 5.21–5.23. As indicated in these figures, the isotherms are found to represent the data closely.

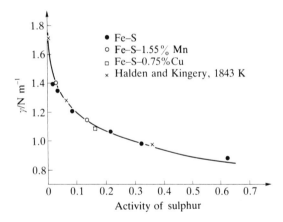

FIG. 5.21. The depression of the surface tension of iron by sulphur, and comparison with the ideal isotherm (5.53) for 1823 K (after Belton 1976).

The effect of alloying on the surface tension of liquid metals in dilute solutions has been compiled by Wilson (1965d) and reviewed by Allen (1972f). Roughly speaking, as pointed out by Baes and Kellogg (1953), the elements most likely to be highly surface active in liquid metals are those of limited solubility in the liquid metal and possessing considerably weaker inter-molecular bonding forces than the metal itself. The elements of the periodic

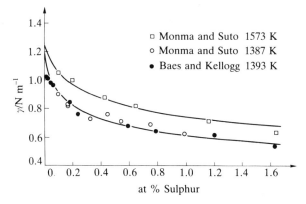

FIG. 5.22. Comparison of the isotherms (5.57) and (5.59) for 1393 and 1573 K with the data for the depression of the surface tension of copper by sulphur (after Belton 1976).

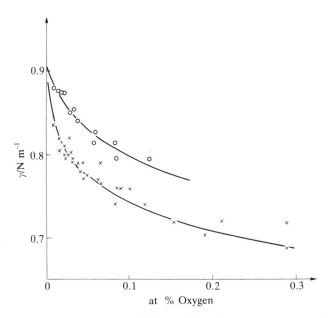

FIG. 5.23. Comparison of the isotherms (5.60) and (5.61) with the measurements of Bernard and Lupis for the effect of oxygen on silver. Upper line and open circles, 1381 K; lower line and crosses, 1253 K (after Belton 1976).

Groups V and VI are apt to be strongly surface active in liquid metals. On the other hand, in dilute solutions a high-surface-tension solute (in general, a high-melting-point metal) is expected to have little effect on the surface tension of solutions. These considerations appear to be substantiated by the results of experimental investigations. However, as described previously, common trace impurities can produce considerable effects on surface tension. Systematic investigations into the effects of solute additions on the surface tension of carefully purified liquid metals are therefore highly desirable.

5.8. EQUATIONS FOR THE SURFACE TENSION OF BINARY LIQUID MIXTURES

The surface tensions of most binary liquid mixtures exhibit negative deviations from the proportional mathematical addition of the pure components' surface tensions. This is often the result of the liquid surface becoming enriched with the component of lower surface tension.

Several theoretical equations describing the surface tension of binary liquid mixtures have been proposed. However, none is completely satisfactory, owing to a lack of information on the structures of (binary) liquid surfaces. Furthermore, no empirical relation has been presented for the surface tension of binary liquid alloys.

By extending Gibbs treatment on the surface tension of mixtures, Butler (1932) deduced a set of equations of the form

$$\begin{aligned}
\gamma_M &= \gamma_1 + \frac{RT}{A_1} \ln \frac{a_1^s}{a_1} \\
&= \gamma_2 + \frac{RT}{A_2} \ln \frac{a_2^s}{a_2} \\
&= \ldots
\end{aligned} \tag{5.62}$$

in which γ_M and $\gamma_1, \gamma_2, \ldots$, represent the surface tension of the mixture and its components 1, 2, $\ldots$, respectively; $A_1, A_2, \ldots$, are the respective areas of components 1, 2, $\ldots$, occupied in monolayers, i.e. molar surface area; a_1, $a_2, \ldots$, are Raoultian or mole fraction activities of the components in the bulk of the mixture referred to the pure bulk components as the standard states; and $a_1^s, a_2^s, \ldots$, are the component's activities in the surface monolayer in which the surface monolayers of the pure components are taken as the standard states.

In deriving eqn (5.62), the essential assumption made is that the difference in composition of the surface from that of the bulk is entirely restricted to the first layer of molecules. For ideal (or perfect) binary mixtures, eqn (5.62) may be written:

$$\gamma_M = \gamma_1 + \frac{RT}{A_1} \ln \frac{x_1^s}{x_1}$$

$$= \gamma_2 + \frac{RT}{A_2} \ln \frac{x_2^s}{x_2} \qquad (5.63)$$

in which x_1, x_2 and x_1^s, x_2^s are the mole fractions in the bulk and in the surface monolayer, respectively.

Guggenheim (1945), using a statistical mechanical approach (i.e. the method of grand partition function), derived equations for the surface tension of binary solutions on the assumption that the difference in composition of surface from the bulk is confined to a unimolecular layer. Guggenheim's equations can be expressed in the following form: for the surface tension of a perfect binary solution,

$$\exp\left(-\frac{\gamma_M A}{RT}\right) = x_1 \exp\left(-\frac{\gamma_1 A}{RT}\right) + x_2 \exp\left(-\frac{\gamma_2 A}{RT}\right) \qquad (5.64)$$

and for the surface tension of a binary regular solution,

$$\gamma_M = \gamma_1 + \frac{RT}{A} \ln \frac{x_1^s}{x_1} + \frac{W}{A} l\{(x_2^s)^2 - x_2^2\} - \frac{W}{A} m x_2^2$$

$$= \gamma_2 + \frac{RT}{A} \ln \frac{x_2^s}{x_2} + \frac{W}{A} l\{(x_1^s)^2 - x_1^2\} - \frac{W}{A} m x_1^2 \qquad (5.65)$$

where A is the molar surface area[†], $W = H^E/x_1 x_2$ (H^E is the enthalpy of mixing), l and m are the fractions of the total of next-neighbour contacts made by a molecule (any molecule in the surface layer) within its own layer and with molecules in the next layer ($l + 2m = 1$).[‡]

Bernard and Lupis (1971) have determined the surface tension of silver–gold alloys and have demonstrated that, as shown in Fig. 5.24, the experimental results agree very well with those calculated by eqn (5.64).

Belton and Evans (1945) have also proposed equations based on a statistical method for calculating the surface tensions of binary liquid systems which form perfect solutions.

Guggenheim's treatment of the surface tension for a binary regular solution has been modified by Hoar and Melford (1957) and by Monma and Suto (1961). The modified expressions are (Hoar and Melford 1957).

[†] In Guggenheim's approach, the molar areas A_1 and A_2 are assumed to be equal, i.e. $A_1 = A_2 = A$.
[‡] In a simple cubic lattice, $l = 2/3$ and $m = 1/6$, while in a close packed lattice, $l = 1/2$ and $m = 1/4$.

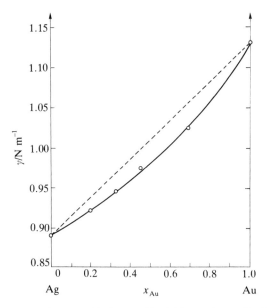

FIG. 5.24. Surface tension of silver–gold alloys at 1381 K. The continuous line is predicted by the 'perfect solution' model (after Bernard and Lupis 1971).

$$\gamma_{\mathrm{M}} = \gamma_1 + \frac{RT}{A_1} \ln \frac{x_1^s}{x_1} + \frac{W}{A_1} \{l'(x_2^s)^2 - x_2^2\}$$

$$= \gamma_2 + \frac{RT}{A_2} \ln \frac{x_2^s}{x_2} + \frac{W}{A_2} \{l'(x_1^s)^2 - x_1^2\} \qquad (5.66)$$

in which l' is the fractional factor. According to their investigations, the value of l' lies in the range 0.5–0.75 for binary liquid alloys.

Although eqn (5.66) gives relatively good agreement with the experimental data for the surface tensions of binary lead–tin and lead–indium alloys (Hoar and Melford 1957), its performance has been rather poor for metallic systems which present large deviations from ideality, and for compounds in the solid state (Bernard and Lupis 1971).

6

VISCOSITY

6.1. INTRODUCTION

When the gradient of a property such as temperature exists in a liquid, a transport process occurs in that liquid. The transport process is a non-equilibrium, or irreversible, process, in which the property (e.g. temperature) can change with time if spatial variations of the property exist within the liquid. The well-known transport processes of momentum, mass, and energy involve viscosity, diffusion, and thermal conduction, respectively. This chapter is devoted to a discussion of the viscosity of metallic liquids. Chapters 7 and 8 cover diffusion and electrical and thermal conductivities, respectively.

Consider, therefore, momentum transport processes that occur when an incompressible liquid is subjected to a uniform shear stress. A velocity gradient is set up perpendicular to the direction of the applied stress, as a result of the fluid's resistance to the applied motion. This resistance is known as a viscous force. Thus, when adjacent parts of a liquid move at different velocities (i.e. a velocity gradient is present), viscous forces act so as to cause the slower-moving regions to move more rapidly, and the faster-moving ones to move more slowly. Thus, viscosity is a physical property which only manifests itself when a relative motion between different layers of fluid is set up.

We now consider the problem of viscosity from a microscopic viewpoint. Although the nearest-neighbour distances and coordination numbers in the liquid state at or near the melting temperature are closely similar to those in the solid state, the dynamic behaviour of atoms in the two states is entirely different. From the microscopic point of view, the most characteristic feature of a liquid is the high mobility of its individual atoms. However, the motions of atoms through a liquid are impeded by frictional forces set up by their nearest neighbours. Viscosity is, therefore, also a measure of the friction among atoms. Consequently, a liquid's viscosity is of great interest in both the technology and theory of liquid metal behaviour.

From a practical standpoint, viscosity plays an important role as a key to solve quantitatively problems in fluid flow behaviour as well as those related to the kinetics of reactions in metallurgical processes. For example, a liquid metal's viscosity is a main factor dominating the rise of small gas bubbles and non-metallic inclusions through it. Similarly, useful information on the rates of slag (flux)/metal reactions or the rates of transfer of impurity element from metal to slag can be obtained by continuous monitoring of a slag's viscosity during reaction, and attendant composition changes.

From a standpoint of theory, other dynamic properties of liquids, e.g.

diffusion, involve viscosity as an essential quantity. Several theoreticians have proposed equations for the viscosity of liquids based on statistical kinetic theories or non-equilibrium statistical mechanics. Consequently, scientists also have a keen interest in the viscosity of liquid metals since they represent the simplest forms of monatomic substances.

Numerous experimental measurements of liquid metal viscosities have been made over the last hundred years or more. Even so, accurate and reliable data are still not in abundance. Fairly large discrepancies exist between experimental viscosities obtained for some liquid metals, particularly iron, aluminium, and zinc. The reason for these discrepancies was attributed to the high reactivity of metallic liquids and the technical difficulty of taking precise measurements at elevated temperatures. However, recent investigations by Iida and co-workers (Iida, Satoh, Ishiura, Ishiguro, and Morita, 1980; Iida, Kumada, Washio, and Morita, 1980) indicate that a part of these large discrepancies is due to the lack of a rigorous formula for calculating viscosities.

6.2. METHODS OF VISCOSITY MEASUREMENT

The definition of a coefficient of viscosity, dynamic viscosity (shear viscosity), or simply 'viscosity' is based on the following mathematical expression by Sir Isaac Newton:

$$\tau = \mu \frac{dv}{dz} \tag{6.1}†$$

where τ is the force exerted by the fluid per unit area of a plane parallel to the x direction of motion when the velocity v is increasing with distance z measured normal to the plane, at the rate dv/dz. Thus, viscosity represents the constant of proportionality. Equation (6.1) is known as Newton's law of viscosity. If a liquid obeys this equation (6.1), i.e. the shearing force τ is proportional to the velocity gradient dv/dz, it is known as a Newtonian liquid. All liquid metals are believed to be Newtonian‡.

Incidentally, it should be kept in mind that the definition of viscosity by eqn (6.1) only holds for laminar or streamline flow. Turbulent viscosities are often defined by a similar equation and apply when flow is turbulent. The latter are typically 10^4 to 10^6 times greater than their laminar counterparts.

The reciprocal of viscosity is known as fluidity. The ratio of viscosity to density (i.e. $v = \mu/\rho$) is known as the fluid's kinematic viscosity v. This is an important physical quantity in fluid mechanics. The dimensions of kinematic

---

† Newton's law of viscosity is sometimes written as

$$\tau_{zx} = -\mu \frac{dv_x}{dz}.$$

This states that the shear force per unit area τ_{zx} is proportional to the *negative* of the local velocity gradient dv_x/dz (see, e.g. Bird, Steward, and Lightfoot 1960).

‡ In general, liquids with comparatively small molecules are considered to be Newtonian liquids.

viscosity $[L^2T^{-1}]$ are equivalent to those of diffusivity, so that v can also be considered to represent the transverse diffusion of momentum down a velocity gradient.

A variety of methods exist for measuring viscosities of liquids. However, suitable techniques for determination of the viscosity of liquid metals and alloys are restricted to the following on account of their low viscosities[†], their chemical reactivity, and their generally high melting points. They are:

    (a) The Capillary Method,
    (b) The Oscillating-vessel Method,
    (c) The Rotational Method,
    (d) The Oscillating-plate Method.

6.2.1. Capillary method

When a liquid is made to flow through a capillary tube under a given pressure, the time required for a definite volume of the liquid to be discharged depends on the viscosity of the liquid. In the capillary method, viscosities can be determined by measuring efflux times of liquid samples through the capillary tube. In practice, as can be easily understood from Fig. 6.1, a definite volume of liquid is introduced into the reservoir bulb, and then sucked into the measuring bulb. The efflux time t for the meniscus to pass between the fiducial marks m_1 and m_2 is then observed. The meniscus can be observed directly by eye, or indirectly using an electrical method. The relation between viscosity and efflux time is then given by the modified Poiseuille equation or Hagen–Poiseuille equation as follows:

$$\mu = \frac{\pi r^4 \rho g \bar{h} t}{8V(l + nr)} - \frac{m\rho V}{8\pi(l + nr)t} \tag{6.2}$$

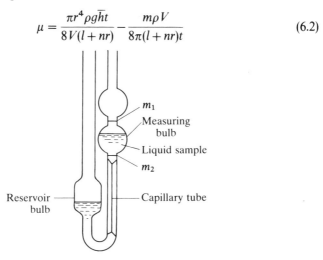

FIG. 6.1. Capillary viscometer.

[†] Liquid metals and alloys are particularly notable for their low kinematic viscosities.

where r and l are the radius and length of the capillary, respectively, $\bar{h}$ is the effective height of the column of liquid (the difference in height between the levels of the liquid in the two tubes), ρ is the liquid density, V is the volume discharged in time t (the volume of the measuring bulb), and m and n are constants ($m = 1.1$–1.2; $n = 0$–0.6). In eqn (6.2), the second term is the kinetic-energy correction which is important in determining the vicosities of liquid metals, and the term nr is called the end-correction.

Relative, rather than absolute, viscosity measurements have usually been made using this technique. This is especially true since the experimental procedures are simple, and any errors incidental to the measurement of dimensions are thereby avoided. For a viscometer in which r, l, $\bar{h}$, and V are fixed, eqn (6.2) reduces to

$$\frac{\mu}{\rho} = v = C_1 t - \frac{C_2}{t} \qquad (6.3)$$

$$C_1 = \frac{\pi r^4 g \bar{h}}{8V(l + nr)}, \; C_2 = \frac{mV}{8\pi(l + nr)}$$

where C_1 and C_2 are constants. The values of C_1 and C_2 are easily evaluated using Standard Viscosity Samples.

In determining the viscosities of metallic liquids by the capillary method, an especially fine and long-bore tube (in general, $r < 0.15$–0.2 mm, $l > 70$–80 mm) is needed so as to satisfy the condition of a low Reynolds number for ensuring laminar flow. Material limitations for fine, long-bore tubes, and the resulting need for particularly clean liquid metal specimens, represent two of the disadvantages of this technique. Only capillary tubes of heat-resisting glass and quartz glass have been used to date for liquid metal containment. The method has, therefore, been applied to determinations of viscosity of metals with melting points below 1400–1500 K.

From the metallurgical standpoint, the problem of possible contamination of liquid metal samples is of critical importance. Blockage of the capillary tube by minute non-metallic inclusions and bubbles has a direct and pronounced effect upon the efflux time. Therefore, clean liquid metal perfectly free from any contamination is indispensable.

However, metrologically, the capillary method was already established and all relevant experimental corrections were known. Careful measurements using this technique can provide reliable viscosity data for low-melting liquid metals, and give an accuracy better than ± 0.5 per cent.

Examples of viscosity measurements obtained via the capillary method are: gallium (Spells 1936; Menz and Sauerwald 1966; Iida, Morita, and Takeuchi 1975); sodium and potassium (Ewing, Grand, and Miller 1951; Kitajima, Saito, and Shimoji 1976); mercury, indium, bismuth, tin, lead, zinc, cadmium, antimony, silver, and copper (Menz and Sauerwald 1966; Iida, Morita, and Takeuchi 1975).

6.2.2. Oscillating-vessel method

When a liquid is placed in a vessel hung by a tortional suspension, and the vessel is set in oscillation about a vertical axis, the resulting motion is gradually damped on account of frictional energy absorption and dissipation within the liquid. The viscosity of the liquid sample can be calculated by observing the decrement and the time period of the oscillations. This is the principle of the oscillating-vessel method. Diagrams for the oscillation system of measurement are shown in Figs. 6.2 and 6.3.

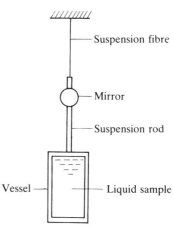

Fig. 6.2 Oscillating-vessel viscometer.

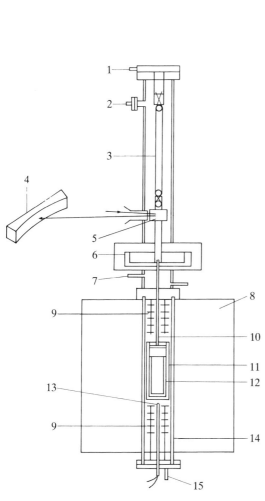

1	Oscillation starter
2	Gas inlet
3	Bifile suspension
4	Camera
5	Mirror
6	Inertia ring
7	Water cooling jacket
8	Mo-resistance furnace
9	Radiation baffles
10	Suspension rod
11	Mo vessel
12	Alumina vessel
13	Thermocouple
14	Alumina tube
15	Gas outlet

FIG. 6.3. Schematic diagram of the oscillating-vessel (or the oscillating-crucible) viscometer (see e.g. Morita, Ogino, Iba, Maehana, and Adachi 1970).

In measuring viscosity at elevated temperatures, this method is the most important and has been most frequently used. Its practical advantages are that the apparatus, and in particular the shape of vessel or crucible, is simple, and a closed vessel can be used. Similarly, the decrement (i.e. the amplitude) and the periodic time of the oscillations can be measured with great accuracy, and the length of the uniform-temperature zone is relatively short (the dimensions of vessels used are usually: high 50–125 mm, internal diameter 20–50 mm). The main disadvantage of this technique is that a rigorous analytical formula for calculating viscosity from the observed decrement and period of the oscillations is lacking owing to the mathematical difficulty of solving the differential equation of motion for this oscillating system. A number of theoretical and experimental investigations has been carried out and several analytical equations proposed for calculating viscosity from experimental data. Individual workers have assessed the errors in measuring liquid metal viscosities, and conclude that they range between 1–5 per cent. However, discrepancies as large as 30–50 per cent exist between the experimental results of different workers (for example, see Fig. 6.17). Iida, Satoh, Ishiura, Ishiguro, and Morita (1980) have made a detailed experimental investigation of this problem. The essential details of their work are described in Subsection 6.3.2.

6.2.3. Rotational method

Consider now a liquid filling the space between two coaxial cylinders (or spheres) as shown in Fig. 6.4. When the outer cylinder rotates with a constant angular velocity, and the inner cylinder remains fixed, the viscous liquid exerts

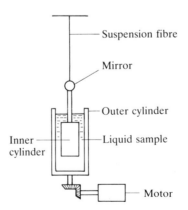

FIG. 6.4. Rotational viscometer.

a revolving force on the inner cylinder. If the inner cylinder is suspended by a fibre, the revolving force, i.e. the torque, can be evaluated by measuring the angular displacement of the fibre. Viscosities can be calculated from observation of the torquing source.

A mathematical analysis of this system is simple. Various types of viscometers based on the principle of the rotational technique exist: rotation of sphere, rotation of disk, and rotation of cylinder (outer cylinder rotated, inner cylinder suspended; outer cylinder fixed, inner cylinder rotated).

However, the application of the rotational method to metallic liquids is technically difficult. For example, in the case of low-viscosity liquids, the clearance allowed between the two cylinders must be very small, and the rotating and stationary parts, therefore, truly coaxial. Only a few investigators (e.g. Jones and Bartlett 1952–3; Krieger and Trenkler 1971) have made use of this technique.

6.2.4. Oscillating-plate method

When a flat plate executing linear oscillations is immersed in a liquid as shown in Fig. 6.5, its motions are impeded by the retarding force which the viscous liquid exerts on the oscillating-plate. If the plate is now vibrating in the liquid with a constant driving force, the amplitude of motion of the plate is reduced to a degree dependent on the viscosity of the liquid. The oscillating-plate method is based on measurements of the amplitudes of plate oscillations in vacuum (or air) and in a liquid sample. The relation between viscosity and amplitudes is expressed in the following form:

$$\rho\mu = K_0 \left(\frac{f_a E_a}{f E} - 1 \right)^2 \tag{6.4}$$

$$K_0 \equiv \frac{R_M^2}{\pi f A^2}$$

where f_a and f are the resonant frequencies for the plate in air and in liquid, respectively, E_a and E are the resonant amplitudes of plate oscillation in air and in liquid, respectively, K_0 is a constant of the apparatus, R_M is the real component of the mechanical impedance, and A is the (effective) area of the plate.

Except for liquids having high values of $\rho\mu$, f may, with good approximation, be assumed to be equal to f_a (Woodward 1953). Equation (6.4) then becomes

$$\rho\mu = K_0 \left(\frac{E_a}{E} - 1 \right)^2 = K_0 \left(\frac{1}{E/E_a} - 1 \right)^2 \tag{6.5}$$

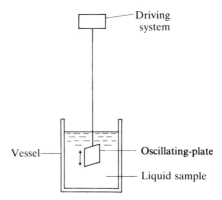

F IG. 6.5. Oscillating-plate viscometer.

The apparatus' constant K_0 is determined experimentally by the use of Standard Viscosity Samples.

The advantages of this method are that it allows instantaneous and continuous indications of the product $\rho\mu$ over a wide range. In addition, its construction and operation are comparatively simple. The technique has been applied to determine the viscosity of liquid iron by Arsentiev, Vinogradov, and Lisichkii (1974). However, the oscillating-plate method is unsuitable for measurements with low-viscosity liquid metal since the thin oscillating-plate of large area must only be vibrated slowly within the liquid.

An application of this technique to metallurgical problems has, however, been made by Iida, Morita, and Chikazawa (1978), and Morita, Iida, Kawamoto, and Mōri (1984). The viscosity apparatus employed is illustrated in Fig. 6.6. The values of viscosity of liquid Na_2CO_3–Li_2CO_3 binary mixtures during chemical reaction with Sn–P alloys were instantaneously and continuously measured as a function of time using the new oscillating-plate viscometer (of resonant frequency 22.5 Hz). The amplitudes of the plate were observed precisely using a non-contact electro-optical measurement system, and recorded automatically. Figure 6.7 provides a typical example of experimental results[†]. In this experiment, a liquid Sn–2.0wt.%P alloy (100 g) was added into liquid Na_2CO_3–50 mol% Li_2CO_3 flux (420 g) at 984 ± 8 K. As can be seen from Fig. 6.7, values for $\rho\mu$ for the flux first increased very rapidly and then decreased rapidly up to about 700 s. Subsequently, values of $\rho\mu$ increased slowly, in an oscillatory way, before levelling off at a roughly constant value.

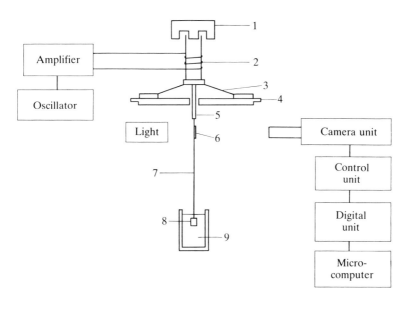

1 Magnet   2 Moving coil   3 Damper

4 Water jacket   5 Connecting rod   6 Target

7 Stem   8 Oscillating-plate

9 Liquid sample

FIG. 6.6. Schematic diagram of an oscillating-plate viscometer (after Morita, Iida, Kawamoto, and Mōri 1984).

The analysis indicated that the P content of the Sn–P alloy had been reduced from 2.0 wt.% to 0.03 wt.% by this flux (slag)/metal reaction. From the experimental results for the viscosity–time curve, i.e. changes in viscosity of the flux occurring with time due to chemical reaction between the liquid Na_2CO_3–50 mol% Li_2CO_3 mixture and liquid Sn–2.0 wt.% P alloy, it can be estimated that this reaction was practically completed during the first 700 s. In the case of reactions between liquid Na_2CO_3–50 mol% Li_2CO_3 flux (50 g) and liquid Sn–4.1 wt.% P alloy (1 g), the P content of the Sn–P alloy was reduced to 0.01 wt.% during the first 60 s.

[†] In these experiments, crucibles of sintered alumina, and square oscillating-plates of high-purity alumina (length of side, 30 mm; thickness, 0.8 mm), were employed. No reactions were observed between the alumina and the liquid samples.

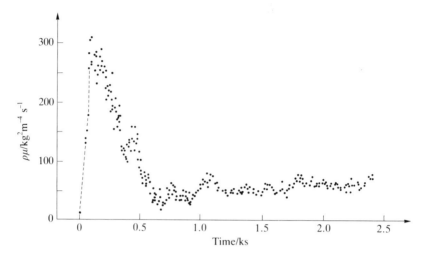

FIG. 6.7. Changes in $\rho\mu$ during chemical (dephosphorization) reaction (after Kijima 1983).

These investigations demonstrate that the physical properties of metals play an important role in reaction velocity. When the densities of reaction products are lower than the density of the liquid flux, the flux/metal reaction goes well since the liquid flux is stirred by the rise of the products (rise velocities depend upon the viscosity of the flux). On the other hand, when the densities of products are higher than the density of liquid flux, the reaction is depressed, with the products piling up on the flux/metal interface. Thus, a knowledge of physical properties of liquids is of critical importance in liquid metal processing operations or extractive metallurgy.

Methods of viscosity measurement for liquid metals are well reviewed in several papers (e.g. Thresh 1962).

6.3. VISCOSITY DETERMINATION BY THE OSCILLATING-VESSEL METHOD (AN EXPERIMENTAL INVESTIGATION OF WORKING FORMULAE)

6.3.1. Working formulae for viscosity determination by the oscillating-vessel method

Although there are various techniques for viscosity measurement, the oscillating-vessel method has been used most extensively for liquid metals and alloys. Unfortunately, the calculation of the viscosity of a liquid from the observed logarithmic decrement and period of the oscillations is extremely complicated. A major reason for large discrepancies among experimental viscosity data is the result of approximations in the working formulae used to connect the observed damping of oscillations and dimensions of the apparatus with the viscosity of the liquid.

The following approximate working formulae have frequently been employed for determinations of viscosity of liquid metals and alloys by the oscillating-cylindrical-vessel method.

6.3.1.1. Knappwost's equation

Knappwost (1948, 1952) has proposed a semi-empirical equation for the calculation of viscosity from the measured logarithmic decrement δ and time period T of a system filled with a liquid sample for a cylindrical vessel of small aspect ratio (i.e. height to radius).

$$\delta T^{3/2} = K(\rho\mu)^{\frac{1}{2}} \tag{6.6}$$

in which K is an apparatus constant. The value of K is determined, previous to viscosity measurements, by using Viscometer-Calibrating Liquids (e.g. mercury, tin, lead) or Standard Viscosity Samples. Knappwost's formula has often been used for relative viscosity determinations of liquid metals and alloys because of its simplicity.

6.3.1.2. Roscoe's equation

Roscoe (1958; Roscoe and Bainbridge 1958) has derived working formulae for absolute viscosity determinations. For a cylindrical vessel, Roscoe's equation is given by

$$\mu = \left(\frac{I\delta}{\pi R^3 H Z} \right)^2 \frac{1}{\pi \rho T} \tag{6.7}$$

where

$$Z = \left(1 + \frac{R}{4H} \right) a_0 - \left(\frac{3}{2} + \frac{4R}{\pi H} \right) \frac{1}{p} + \left(\frac{3}{8} + \frac{9R}{4H} \right) \frac{a_2}{2p^2}$$

$$p = \left(\frac{\pi \rho}{\mu T} \right)^{1/2} R$$

$$a_0 = 1 - \frac{1}{2}\Delta - \frac{3}{8}\Delta^2$$

$$a_2 = 1 + \frac{1}{2}\Delta + \frac{1}{8}\Delta^2$$

$$\Delta = \frac{\delta}{2\pi}$$

I is the moment of inertia of the suspended system, R is the radius of the vessel, and H is the height of the liquid sample in the vessel.

In general, the value of Δ is small compared to 1 since δ is of the order of 10^{-2}–10^{-3}, and so eqn (6.7) may be assumed, to good approximation, to be

$$\frac{\delta}{\rho} = A\left(\frac{\mu}{\rho}\right)^{\frac{1}{2}} - B\left(\frac{\mu}{\rho}\right) + C\left(\frac{\mu}{\rho}\right)^{\frac{3}{2}} \tag{6.8}$$

where

$$A = \frac{\pi^{\frac{3}{2}}}{I}\left(1 + \frac{R}{4H}\right)HR^3T^{\frac{1}{2}}$$

$$B = \frac{\pi}{I}\left(\frac{3}{2} + \frac{4R}{\pi H}\right)HR^2T$$

$$C = \frac{\pi^{\frac{1}{2}}}{2I}\left(\frac{3}{8} + \frac{9R}{4H}\right)HRT^{\frac{3}{2}}$$

Roscoe's absolute formula is considered to provide remarkably accurate viscosity values (e.g. Thresh 1965).

Incidentally, the experimentally determined logarithmic decrement δ is predominantly due to the viscous force of the liquid sample, but it also contains a damping δ_0 arising from internal friction of the suspension fibre and also the protective gas atmosphere (e.g. helium, argon, hydrogen). This damping can be obtained through observation of the logarithmic decrement of the empty container. In calculating viscosities from eqns (6.6)–(6.8), a correction for these effects must be made by subtracting δ_0 from δ (Thresh 1965; Rothwell 1961–2).

6.3.1.3. Shvidkovskii's equation

According to Shvidkovskii, if a parameter $\alpha = R(2\pi/\nu T)^{\frac{1}{2}}$ is greater than 10, and H is greater than or equal to $1.85R$, the working formula connecting the observed data and the dimensions of the apparatus with the kinematic viscosity of the liquid specimen is given (Vertman and Samarin 1969) by

$$\nu = \frac{I^2(\delta - T\delta_0/T_0)^2}{\pi(MR)^2TW^2} \tag{6.9}$$

where

$$W = 1 - \frac{3}{2}\Delta - \frac{3}{8}\Delta^2 - a + (b - c\Delta)\frac{2nR}{H}$$

M represents the mass of the liquid specimen; a, b, and c are constants tabulated as a function of $(2\pi R^2/\nu^* T)$ (ν^* is the value of ν when $W = 1$); n is the number of planes contacted horizontally with the liquid specimen (i.e., in the case of a vessel having its lower end closed and its upper surface free, $n = 1$, if the vessel encloses the fluid top and bottom, $n = 2$); while the subscript 0 refers to an empty vessel.

Shvidkovskii's equation has been used exclusively for absolute or relative viscosity determinations in the USSR. (e.g. Samarin 1962).

Hopkins and Toye (1950), and Toye and Jones (1958) have also presented working formulae for the calculation of viscosity from measured damping data and dimensions of apparatus.

Andrade and Chiong (1936) proposed a formula for absolute viscosity determinations in the case of the oscillating-sphere method, and made careful measurements of the viscosities of the liquid alkali metals (Chiong 1936; Andrade and Dobbs 1952). Although their working formula for a spherical vessel appears to provide accurate and reliable viscosity data, their technique is not popular because of the limitation of materials for the construction of spherical vessels.

6.3.2. Experimental investigations of working formulae

Iida and his co-workers (Iida, Satoh *et al.* 1980; Iida, Kumada *et al.* 1980; Morita and Iida 1982) have examined experimentally the validity of the above-mentioned formulae. The samples used in these investigations were mercury, indium, tin, bismuth, lead, and copper (vacuum melted, 99.99–99.999 % in purity) whose viscosities had been determined precisely using capillary viscometers[†]. The logarithmic decrements, i.e. the amplitudes, and time periods of the oscillations for these samples were measured carefully using two types of oscillating-vessel viscometer. The experimental conditions employed in this work are given in Table 6.1.

The results obtained are as follows.

(1) In Knappwost's formula, the value of K is assumed to be a constant. However, as pointed out by Kleinschmit and Grothe (1970), the apparatus constant K can be expressed in terms of kinematic viscosity. By comparing eqn (6.6) with eqn (6.8), we have

$$K = T^{\frac{3}{2}}\left\{A - B\left(\frac{\mu}{\rho}\right)^{\frac{1}{2}} + C\left(\frac{\mu}{\rho}\right)\right\} \qquad (6.10)[‡]$$

Figures 6.8 and 6.9 show the apparatus constants as a function of kinematic viscosity. For these figures, the values of K for each plot were calculated from

---

[†] In determinations of the viscosity of metallic liquids, these metals are frequently employed as Viscometer-Calibrating Liquids.

[‡] The variations of time periods T with temperature are negligibly small. In addition, if the moment of inertia of the whole suspended system is large, the differences in time periods between liquid metals are also negligibly small.

TABLE 6.1. *Experimental conditions*

	Viscometer No. 1	Viscometer No. 2
Suspension system	unifiler, ϕ 0.3 mm Mo (inverse suspension type)	bifiler, ϕ 0.3 mm Mo (suspension type)
Vessel	∘ graphite ∘ ϕ 20 mm × H 30 mm (H is height of liquid) ∘ closed at upper and lower ends ∘ possible to use the same vessel repeatedly	∘ recrystallized alumina ∘ ϕ 20 mm × 85 mm (H is height of liquid) ∘ upper surface of liquid being free ∘ impossible to use the same vessel repeatedly
Temperature range	467–913 K	948–1953 K
Atmosphere	He (0.1 MPa)	He (0.1 MPa)

Knappwost's formula by eqn (6.6) and the values of the kinematic viscosity were taken from Iida *et al.* (1975; Takeuchi, Morita, and Iida 1971). As shown in Figs. 6.8 and 6.9, the values of the apparatus constant seem to depend not only on kinematic viscosities but also on the kinds of metal used. In conclusion,

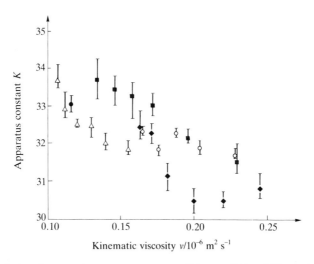

FIG. 6.8. Relations between the apparatus constant K in eqn (6.6) and kinematic viscosity of Viscometer-Calibrating Liquids (VCL) (viscometer No. 1). ○ Lead, ● Mercury, △ Bismuth, ■ Indium, ◆ Tin. The error bars are due to the scatter in measured values. (after Iida, Satoh, Ishiura, Ishiguro, and Morita 1980).

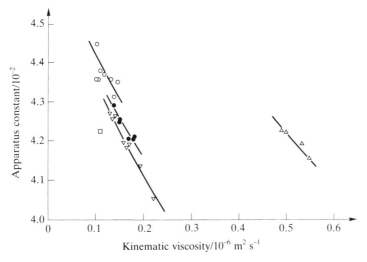

FIG. 6.9. Relations between the apparatus constant K in eqn (6.6) and kinematic viscosity of VCL (viscometer No. 2). □ Mercury, ○ Bismuth, ● Lead, ∇ Tin, △ Copper. (after Morita and Iida 1982).

Knappwost's working formula, in general, gives roughly approximate values for viscosity.

(2) Values for viscosity of liquid metals employed in these studies were calculated from experimental data on the logarithmic decrements and periods of oscillations using eqns (6.6), (6.8) and (6.9). Absolute viscosities were calculated using Roscoe's formula and Shvidkovskii's formula. Relative viscosities were also calculated using Knappwost's formula in which a value for the apparatus constant was determined using mercury at room temperature. Figures 6.10 and 6.11 show computed values for the viscosity of liquid bismuth and iron together with independent reference data determined using the capillary method (Iida *et al.* 1975). Although identical experimental data were therefore used in the calculation of viscosity, one sees that considerable discrepancies exist among not only viscosity values but also temperature dependence (i.e. apparent activation energies). For example, the calculated value for the viscosity of liquid bismuth at 700 K using Roscoe's absolute formula is approximately 20% lower than the result using Knappwost's relative formula.

Figures 6.12 and 6.13 indicate comparisons of the computed values for the viscosity of liquid zinc using Roscoe's absolute equation and Knappwost's relative equation, where the apparatus constant was obtained by using mercury, indium, tin, bismuth, and lead, with the published values of various workers.

From these figures, one finds that calculated values of liquid metal viscosities are considerably dependent on working formulae.

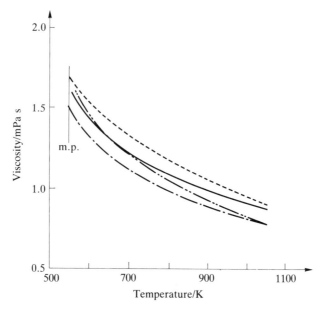

FIG. 6.10. Calculated values for viscosity of liquid bismuth. Working formulae used are:
-------- Knappwost's formula by eqn (6.6);
————-—— Roscoe's formula by eqn (6.8);
————---——Shvidkovskii's formula by eqn (6.9);
———————— measured by a capillary viscometer.
(after Iida and Morita 1980; Morita and Iida 1982).

(3) Although Roscoe's equation might be believed to provide the most accurate values for viscosity, the agreement of viscosity values calculated from Roscoe's formula with the data obtained by the capillary method is not good. For this reason, experimental investigations were made on the reliability of Roscoe's equation. Figure 6.14 shows a result of the determination of end effects. As is obvious from this figure, the values calculated from Roscoe's absolute formula do not coincide with experimental data. Since the logarithmic decrement $(\delta - \delta_0)$ at $H = 0$ comes from the end effect, Roscoe's formula provides insufficient weighting of the end correction. If we now assume that the end effect corresponds to the increase in the height of the liquid sample then, in the case of liquid lead contained in a cylindrical vessel of graphite, the value of the end correction ΔH to be added to Roscoe's absolute formula is about 3 mm according to the experimental result in Fig. 6.14. The value of 3 mm for the end correction (in this investigation, $(H + \Delta H)/H = 1.1$) should not be neglected.

However, the major disagreement in the slopes $((\delta - \delta_0)$ vs. H, in Fig. 6.14) between the calculated values and experimental data may be due to a slipping phenomenon. A possible explanation for the smaller slope exhibited by the

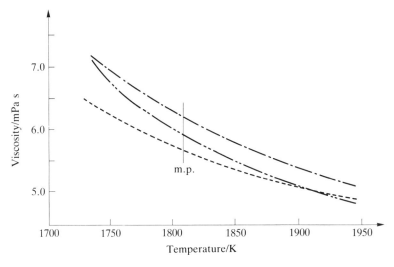

FIG. 6.11. Calculated values for viscosity of liquid iron. Working formulae used are:

-------- Knappwost's formula by eqn (6.6);
———·——— Roscoe's formula by eqn (6.8);
———··———Shvidkovskii's formula by eqn (6.9).
(after Morita and Iida 1982).

experimental data is that the liquid lead did not completely wet the sidewalls of the vessel. The resulting slippage between the liquid interface and adjacent graphite walls would provide smaller damping than anticipated (Thresh 1965; Fisher and Phillips 1954; Ofte and Wittenberg 1963).

(4) In order to add these corrections to Roscoe's absolute formula, a parameter ζ (called correction factor) was, as a matter of convenience, introduced.

$$\frac{\delta - \delta_0}{\rho} = \zeta \left\{ A\left(\frac{\mu}{\rho}\right)^{\frac{1}{2}} - B\left(\frac{\mu}{\rho}\right) + C\left(\frac{\mu}{\rho}\right)^{\frac{3}{2}} \right\} \qquad (6.11)$$

Figure 6.15 indicates the relationship between the ratio of the logarithmic decrement to density $(\delta - \delta_0)/\rho$ and the kinematic viscosity $v = \mu/\rho$. In this figure, the curve $\zeta = 1$ is identical to values calculated from Roscoe's absolute equation. As expected, one finds that Roscoe's equation does not necessarily reproduce the kinematic viscosities of the five Viscometer-Calibrating Liquids with good agreement. If eqn (6.11) is called the corrected Roscoe's formula, a corrected Roscoe's formula employing 1.04 for the correction factor, best reproduces, on average, data for the Viscometer-Calibrating Liquids. Figure 6.16 shows values computed from the corrected form of Roscoe's equation for the viscosity of liquid zinc together with values obtained by other investigators.

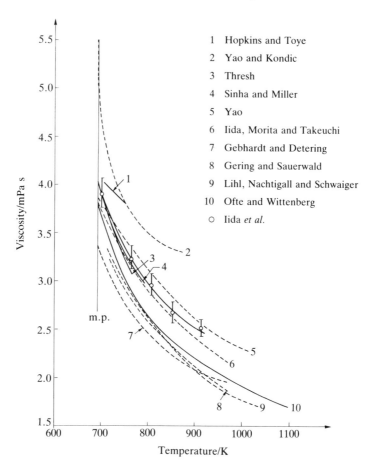

FIG. 6.12. Viscosity of liquid zinc as a function of temperature. Open circles represent values calculated from Roscoe's absolute formula by eqn (6.8). The continuous curves represent values calculated by the absolute method. The bars are due to the probable errors of VCL (after Iida, Satoh, *et al.* 1980).

In calculating the viscosity of liquid zinc, the value of 1.04 was assumed for the correction factor ζ.

The experimental data determined by a number of workers for the viscosity of liquid iron are shown in Fig. 6.17. These viscosity values were measured using oscillating-vessel methods, except for the data of Arsentiev *et al.* (Arsentiev, Vinogradov, and Lisichkii, 1974) and Krieger and Trenkler (1971). In calculating the viscosity, the following formulae were used: the investigators in the USSR, Shvidkovskii's equation; Ogino *et al.* (Ogino, Borgmann, and Frohberg 1973), Frohberg and Cakici (1977), Roscoe's equation; the other

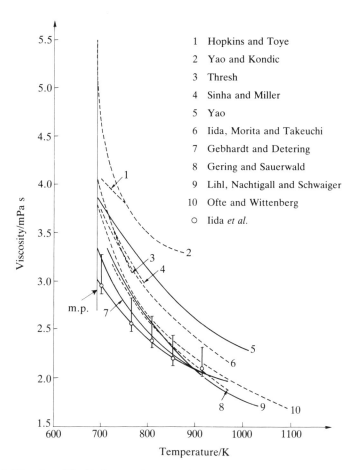

FIG. 6.13. Viscosity of liquid zinc as a function of temperature. Open circles and continuous curves represent values calculated from Knappwost's formula. The bars are due to the probable errors of VCL (after Iida, Satoh, *et al.* 1980).

investigators, Knappwost's equation. As is obvious from Fig. 6.17, workers in the USSR report viscosity values which are higher near the melting point and exhibit greater temperature dependence in comparison with the data of others. This discrepancy can mainly be attributed to the approximate nature of their working formulae.

The viscosities of liquid iron were calculated from the experimental data using the corrected version of Roscoe's equation. A value of 0.96 ± 0.4 used for the correction factor was obtained using tin, bismuth, lead, and copper. Figure 6.18 shows the present data for the viscosity of liquid iron versus temperature.

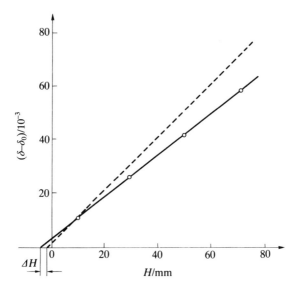

FIG. 6.14. Comparison of the calculated and measured values of end correction for liquid lead in a cylindrical vessel of graphite at 672 K. ----Calculated, O Measured. (after Iida, Kumada, Washio, and Morita 1980; Iida and Morita 1980).

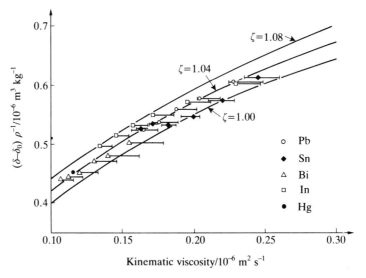

FIG. 6.15. Relation between the ratio of logarithmic decrement to density, and kinematic viscosity, for several liquid metals. The bars indicate the probable errors of VCL (after Iida, Satoh, *et al.* 1980; Iida and Morita 1980).

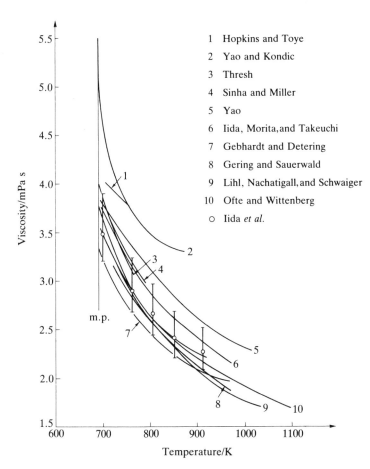

FIG. 6.16. Viscosity of liquid zinc as a function of temperature. Open circles represent values calculated from the corrected Roscoe's formula by eqn (6.11), where the correction factor ζ has been taken to be equal to 1.04 (viscometer No. 1). The bars indicate the range of values calculated by using the correction factor between 1.00 and 1.08. (after Iida, Satoh, et al. 1980).

Incidentally, it should be kept in mind that the best value of the correction factor ζ depends on the construction of the viscosity apparatus, and, especially, on the materials and dimensions of the vessel.

6.4. THEORETICAL EQUATIONS FOR VISCOSITY

Exact expressions for the equilibrium properties of liquids have been formulated on the basis of statistical mechanical theory. If two basic quantities, the pair distribution function $g(r)$ and the pair potential $\phi(r)$, are available, the

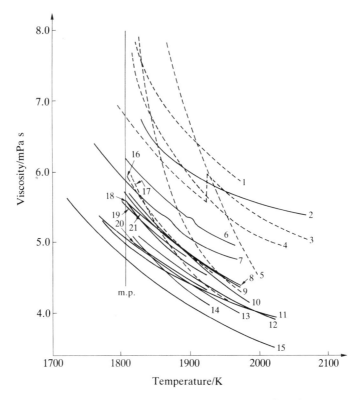

FIG. 6.17. Viscosity of liquid iron, as determined by a number of workers. ----- Values obtained by workers in the USSR. (1) Arsentiev *et al.*, (2) Barfield and Kitchener, (3) Nobohatskii *et al.*, (4) Romanov and Kochegarov, (5) Samarin, (6) Ogino *et al.*, (7) Ogino *et al.*, (8) Nakanishi *et al.*, (9) Vatolin *et al.*, (10) Frohberg and Cakici, (11) Cavalier, (12) Saito and Watanabe, (13) Lucas, (14) Kawai *et al.*, (15) Thiele, (16) Avaliani *et al.*, (17) Wen Li-shi and Arsentiev, (18) Schenck *et al.*, (19) Frohberg and Weber, (20) Narita and Onoe, (21) Krieger and Trenkler. (after Iida and Morita 1980).

thermodynamic properties of liquids can be readily calculated[†]. However, the derivation of calculable, rigorous expressions for the dynamic properties of liquids is extremely difficult since atomic motion in liquids cannot be precisely described as a function of time[‡]. Consequently, various approximate expres-

---

[†] In the statistical mechanical theory of liquids, an assumption is introduced that the interatomic forces and energies in a classical liquid are dominated by the sum of pair interactions. This approach is called the pair theory of liquids.

[‡] In the theory of liquids, the space and the time dependent correlation functions are the most fundamental properties. A complete knowledge of these functions at different densities and temperatures is sufficient to calculate all thermodynamic and transport properties of classical liquids. Unfortunately, there is, at the present time, no practical theory, other than the machine solution of Newton's equation of motion for each atom, for computing time-dependent correlation functions for liquids from first principles.

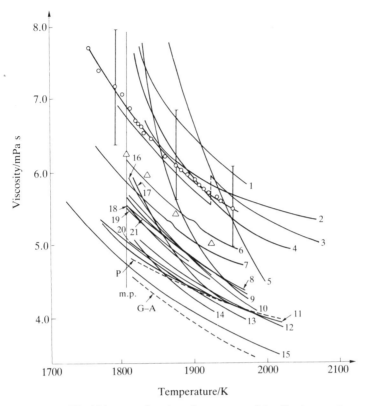

FIG. 6.18. Viscosity of liquid iron as a function of temperature. Identification numbers are the same as those in Fig. 6.17. ○ Measured by Morita and Iida. The bars indicate the range of values calculated by using the correction factor between 0.92 and 1.00 (viscometer No. 2) (after Morita and Iida 1982). ----- Predicted by Pasternak (1972); P. Theory of corresponding states; G–A, Grosse–Andrade method. △ Calculated using eqn (6.40).

sions for transport coefficients have been proposed, which rest on different concepts.

6.4.1. Equations based on the pair distribution function method

(1) Born and Green (1947), using their kinetic theory, have derived an expression for the viscosity of liquids in terms of the pair distribution function and the pair interatomic potential:

$$\mu = \frac{2\pi}{15}\left(\frac{m}{kT}\right)^{\frac{1}{2}} n_0^2 \int_0^\infty g(r)\frac{\partial\phi(r)}{\partial r} r^4 \, \mathrm{d}r \qquad (6.12)$$

in which m is the mass of the atoms (molecules).

Values for viscosity of liquid metals have been calculated through the use of eqn (6.12) by Johnson *et al.* (1964) and others (Shimoji 1967; Waseda and Ohtani 1973; Waseda 1980*f*). In calculating viscosity values, available experimental data for pair distribution functions, and pair potentials (i.e. ion–ion oscillatory potentials) calculated from the Born–Green and the Percus–Yevick integral equations connecting $g(r)$ and $\phi(r)$ (e.g. Egelstaff 1967*a*), were used. As shown in Table 6.2, values calculated from the Born–Green integral equation appear to coincide well with experimental data, with the exception of a few metals. The temperature dependence of viscosity is also semi-quantitatively correct. From a theoretical point of view, however, the validity of eqn (6.12) and the accuracy of the calculations for the pair potentials need thorough investigation.

TABLE 6.2. *Comparison of calculated values for viscosity of liquid metals with experimental data*

Metal	Temperature (K)	Viscosity (mPa s)		
		B–G	P–Y	expt.
Li	453	0.79	1.20	0.59
	473	0.43	1.08	0.55
Na	387	0.70	1.74	0.68
	476	0.59	1.53	~0.40
K	343	0.68	0.94	~0.51
	618	0.44	0.88	0.25
Rb	313	0.78	1.43	0.67
	513	0.53	1.26	~0.32
Cs	303	0.84	1.42	0.63
	573	0.52	1.28	~0.34
Hg	273	1.78	1.24	1.68
	423	1.57	1.09	~1.1
Al	973	0.95	1.35	2.9
	1123	0.88	1.22	~1.3
Pb	623	1.84	1.29	~2.2
	823	1.60	1.18	1.7

Data from Johnson *et al.* (1964).

(2) Rice and co-workers (Rice and Kirkwood 1959; Rice and Allnatt 1961; Allnatt and Rice 1961; Rice and Gray 1965) have also proposed a viscosity equation for dense fluids, in terms of interatomic pair potentials (presented as a rigid core plus an attraction) and pair distribution functions, using their statistical kinetic theory. According to their considerations, the viscosity μ is

divided into three parts: (a) the kinetic contribution μ_K, (b) the part of momentum transfer due to a hard-sphere collision $\mu_\phi(\sigma)$, and (c) a contribution arising from interactions between the soft attractions $\mu_\phi(r > \sigma)$.

The values for the viscosity of five liquid metals, sodium, potassium, zinc, indium, and tin, were calculated using Rice et al.'s formulae and compared with experimental data by Kitajima, Saito, and Shimoji (1976). The comparison between calculated and measured values indicates that the kinetic contribution to shear viscosity is negligibly small $(\mu_K \ll \mu_\phi(\sigma) + \mu_\phi(r > \sigma))$, and that the contribution due to the pair interactions is dominant for the liquid state (thus, the magnitude of $\mu_\phi(r > \sigma)$ represents 70–80 per cent of the total shear viscosity).

Rice and Kirkwood (1959) have proposed an approximate equation for viscosity arising from interactions between soft attractions.

$$\mu_\phi(r > \sigma) = \frac{2\pi m n_0^2}{15\zeta_f} \int_0^\infty r^4 \left\{ \frac{\partial^2 \phi(r)}{\partial r^2} + \frac{4}{r} \frac{\partial \phi(r)}{\partial r} \right\} g(r) \, dr \tag{6.13}$$

in which ζ_f is the friction coefficient.

Nevertheless, the ratio of calculated to measured values is of the order of 0.5, while the agreement between calculation and experiment is still not satisfactory from the viewpoint of metallurgical engineering.

6.4.2. An equation based on a moment method

A state of non-equilibrium can be considered to derive from fluctuations in the equilibrium state. Consequently, transport coefficients are closely related to fluctuations in the number density of atoms during their thermal motion in the liquid state, or to a relaxation phenomenon in density fluctuations (the fluctuation–dissipation theorem) (Rice and Gray 1965; Helfand 1960; Kubo 1957; Kubo, Yokota, and Nakajima 1957).

As yet, no exact theory for calculating space- and time-dependent correlation functions exists. However, a number of useful properties of liquids can be computed using the low-order frequency moment relations which are defined as the coefficients in the short time expansion of the space- and time-dependent correlation functions.

According to Forster, Martin and Yip (1968) the shear viscosity of fluids is represented in terms of the second $(\langle \omega_t^2(Q) \rangle_{av})$ and fourth $(\langle \omega_t^4(Q) \rangle_{av})$ moments of the transverse momentum correlation functions as follows:

$$\mu = mn_0 \lim_{Q \to 0} \frac{(2\pi)^{\frac{1}{2}}}{2} \frac{(\langle \omega_t^2(Q) \rangle_{av}/Q^2)^{\frac{3}{2}}}{[(\langle \omega_t^4(Q) \rangle_{av}/Q^2) - (\langle \omega_t^2(Q) \rangle_{av})^2/Q^2]^{\frac{1}{2}}} \tag{6.14}$$

where ω_t is the transverse angular frequency. In eqn (6.14) the values of the frequency moments, $\langle \omega_t^2(Q) \rangle_{av}/Q^2$ and $\langle \omega_t^4(Q) \rangle_{av}/Q^2$, as $Q \to 0$ can be evaluated from a knowledge of the equilibrium pair distribution functions and interaction potentials.

Values calculated by Kitajima (1976) (see Shimoji 1977a) for liquid sodium and potassium, and by Bansal (1973) for liquid sodium and alluminium, indicate that eqn (6.14) underestimates viscosities by a factor of about 0.5–0.6.

6.4.3. An equation based on a hard-sphere theory

An expression for the shear viscosity of a dense fluid of non-attracting hard spheres due to Longuet-Higgins and Pople (1956) has been discussed by Faber (1972a) and is expressed in terms of the packing fraction η

$$\mu = 3.8 \times 10^{-8} \frac{(MT)^{\frac{1}{2}}}{V^{\frac{2}{3}}} \frac{\eta^{\frac{4}{3}} (1 - \eta/2)}{(1 - \eta)^3} \quad \text{(in Pa s)} \qquad (6.15)$$

Substitution of the value of 0.45 for the packing fraction of liquid metals at their melting points into eqn (6.15) yields the equation for the melting-point viscosity μ_m of a liquid metal

$$\mu_m = 0.61 \times 10^{-7} \frac{(MT_m)^{\frac{1}{2}}}{V_m^{\frac{2}{3}}} \qquad (6.16)$$

This formula underestimates μ_m by a factor of the order of 0.3 (see Subsection 6.5.1, (1) and (5)).

6.5. SEMI-EMPIRICAL (SEMI-THEORETICAL) AND EMPIRICAL EQUATIONS FOR VISCOSITY

Numerous expressions for transport coefficients, i.e. viscosity, diffusivity, have been proposed on the basis of a variety of models for liquids. Well-known examples are the hole theory of Frenkel (1946), the reaction-rate theory of Eyring (e.g. Glasstone, Laidler, and Eyring 1941), the quasi-crystalline theory of Andrade (1934) and the free-volume theory of Cohen and Turnbull (1959). All these model theories are based on phenomenological parameters. In many cases, the parameters must be determined experimentally.

The difficulty inherent in phenomenological analysis or in model theory is that one feature is overemphasized at the expense of another. This difficulty may be overcome only by an adequate basic theoretical treatment, as pointed out by Egelstaff (1967c).

6.5.1. Semi-theoretical equations (model theories)

(1) According to Andrade (1934), the atoms in the liquid state at the melting point may be regarded as executing vibrations about equilibrium positions with random directions and periods just as in the solid state (i.e. Einstein oscillators). Assuming that on the basis of such a quasi-crystalline model the viscosity of a liquid is produced by the transfer of momentum of atomic vibrations from one layer to a neighbouring one, Andrade derived the following equation for the viscosity of simple (monatomic) liquids in the neighbourhood of the melting point:

$$\mu_m = \frac{4}{3}\frac{vm}{a} \tag{6.17}$$

where v is the characteristic frequency of vibration and a is the average distance between the atoms. The numerical factor $\frac{4}{3}$ was roughly estimated.

In calculating the viscosity, Andrade used Lindemann's formula for v and $(V/N_A)^{\frac{1}{3}}$ for a. Thus, eqn (6.17) was rewritten as:

$$\mu_m = 1.6 \times 10^{-4}\frac{(MT_m)^{\frac{1}{2}}}{V_m^{\frac{2}{3}}}, \quad \text{(in mPa s)} \tag{6.18}^{\dagger}$$

where V_m is the atomic volume at the melting point.

Andrade's derivation seems unconvincing from the viewpoint of our present-day theory of liquids. However, eqn (6.18) reproduces the experimental viscosity data for liquid metals at their melting points with comparatively good agreement, as exhibited in Fig. 6.19. Furthermore, Andrade has presented an expression for the temperature variation of viscosity. The expression is

$$\mu v^{\frac{1}{3}} = A\exp\left(\frac{c}{vT}\right) \tag{6.19}$$

in which v is the specific volume, and A and c are constants. Equation (6.19) has been found by many investigators to fit experimental data with a high degree of accuracy. However, as yet, no rigorous theoretical explanation has been given which accounts for the success of this simple equation.

(2) Eyring and co-workers (Eyring, Henderson, Stover, and Eyring 1964; Breitling and Eyring 1972) proposed the theory of significant structure by extending Eyring's early hole theory. According to this model theory, the viscosity for a simple liquid is expressed by

---

[†] According to Andrade and Dobbs (1952), the value of the proportionality constant that gives the best agreement with experimental data for liquid alkali metals is 2.0×10^{-4}. A value of 1.8×10^{-4} is generally used for the proportionality constant in SI units, mPas.

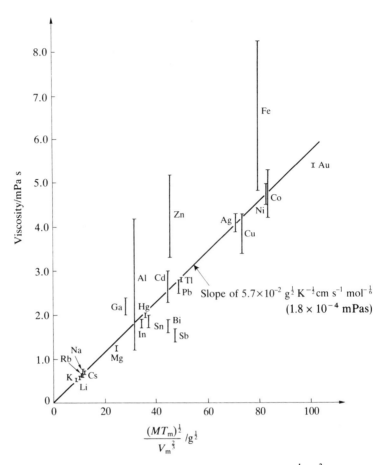

FIG. 6.19. Melting-point viscosities μ_m of various liquid metals vs. $(MT_m)^{\frac{1}{2}} V_m^{-\frac{2}{3}}$. The bars indicate the extremes of modern experimental values for μ_m (after Iida and Morita 1980).

$$\mu = \frac{V_s}{V} \mu_s + \frac{V - V_s}{V} \mu_g \qquad (6.20)$$

where V, V_s, μ_s and μ_g are the molar volumes of liquid and solid, and the viscosities of solid-like and gas-like structures, respectively. In eqn (6.20), a fraction V_s/V of the molecules manifests solid-like behaviour, while the remaining fraction $(V - V_s)/V$ is gas-like[†]. The expressions for μ_s and μ_g are given by

[†] $(V - V_s)$ is the excess volume of the liquid. The excess volume can be considered as dynamic vacancies spread through the liquid.

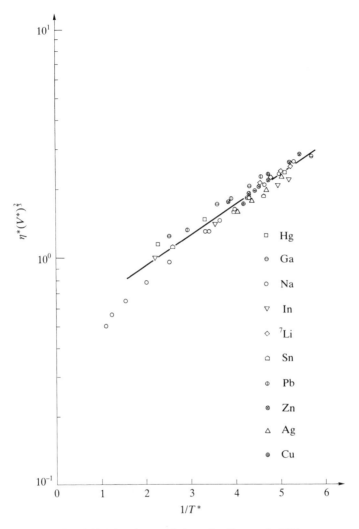

FIG. 6.20. Viscosity correlation (after Pasternak 1972).

$$\mu_s = \frac{6N_A h}{\sqrt{2}Z\kappa} \frac{V}{V_s(V - V_s)} \frac{1}{1 - e^{-\theta/T}} \exp\left\{\frac{a'\Delta_s^g H V_s}{(V - V_s)RT}\right\} \qquad (6.21)$$

$$a' = \frac{(n-1)}{2Z} \frac{(V_m - V_s)^2}{V_m V_s}$$

$$n = \frac{ZV_s}{V_m}$$

$$\mu_g = \frac{2}{3\sigma^2}\left(\frac{mkT}{\pi^3}\right)^{\frac{1}{2}}$$

(6.22)

where h is Planck's constant, Z is the number of nearest neighbours, κ is the transmission coefficient, θ is the Einstein characteristic frequency, $\Delta_s^g H$ is the sublimation energy per mole, and σ is the diameter of the molecule.

Viscosity values for various liquid metals have been computed by Breitling and Eyring (1972) using the above equations. The results show that agreement with experimental data is poor. Their conclusion was that, since all the parameters are fixed, little can be done to calculate better values for liquid metal viscosities until an improved partition function or improved parameters are used.

(3) Macedo and Litovitz (1965) reconsidered the rate-theory approach of Eyring and the free-volume theory of Cohen and Turnbull for liquid viscosities. According to Macedo and Litovitz, expressions for the viscosity of liquid can be formulated by assuming that the following two events must simultaneously occur before a molecule can undergo a diffusive jump: (a) the molecule must attain sufficient energy to break away from its neighbours, and (b) it must have an empty site large enough in which to jump. They have, thus, proposed a hybrid expression for the viscosity of liquids using Eyring's rate theory to calculate (a), and the Cohen and Turnbull approach for (b). This hybrid equation contains both the activation energy and the free volume, and is given by:

$$\mu = A_0 \exp\left(\frac{E_V^*}{RT} + \frac{\gamma_\mu v_0}{v_f}\right)$$

(6.23)

$$A_0 = \left(\frac{RT}{E_V^*}\right)^{\frac{1}{2}} \frac{(2mkT)^{\frac{1}{2}}}{v^{\frac{2}{3}}}$$

where E_V^* is the height of the potential barrier between equilibrium positions, γ_μ the constant between 0.5 and 1 (a numerical factor introduced to correct the overlap of free volume), v_0 the close-packed molecular volume, v_f the average free volume per molecule, and v a quantity roughly equal to the volume of a molecule.

In eqn (6.23), the value of A_0 varies with temperature but is usually far less than the exponential term.

In calculating the viscosity, values of E_V^*, v_0, v_f and v must generally be determined experimentally.

(4) Several workers (Helfand and Rice, 1960; Chapman, 1966; Pasternak, 1972; Wittenberg and DeWitt, 1973; Waseda and Ohtani, 1975) have developed correlations for the transport coefficients of pure liquid metals as a function of temperature using the theory of corresponding states.

Helfand and Rice (1960) have shown that by assuming a pair potential $\phi(r)$ $= \varepsilon\phi^*(r/\sigma_a)$, where ε and σ_a are characteristic energy and distance constants, and ϕ^* is a universal function of r/σ_a, it is possible to apply the theorem of corresponding states for transport properties, i.e. viscosity and diffusion. The viscosity equation based on the corresponding-states theory is given (Pasternak 1972) by

$$\mu = \mu^*(V^*)^{\frac{2}{3}} \frac{(MR\varepsilon/k)^{\frac{1}{2}}}{N_A^{\frac{1}{3}}} \frac{1}{V^{\frac{2}{3}}} \qquad (6.24)$$

where the energy parameter $\varepsilon/k = 5.2\,T_m$.

In Fig. 6.20, values of $\mu^*(V^*)^{2/3}$ for various liquid metals are plotted as a function of the reciprocal of the reduced temperature T^*, where $T^* = Tk/\varepsilon$. Pasternak (1972) has calculated viscosity values for pure liquid iron using eqn (6.24) and Fig. 6.20. Calculated values are in good agreement with Cavalier's experimental data (see Figs. 6.17 and 6.18).

The corresponding-states correlations for the transport coefficients have spreads of ± 20 per cent or more, as a result of the simplicity of the assumptions made for the pair potential.

At the melting temperature, eqn (6.24) becomes similar to Andrade's formula for melting-point viscosity.

(5) From a practical point of view, Andrade's formula for the viscosity of monatomic liquids is worthy of further comment for the following reasons. The Andrade formula is expressed in terms of well-known parameters, M, T_m, and V_m, and gives relatively good agreement with experimental data. At the melting temperatures, other expressions for viscosity (e.g. the Born–Green equation, the equation based on a hard-sphere theory, the equation based on the corresponding-states theory) closely resemble his equation. In addition, Osida (1939) has applied Andrade's considerations of viscosity to the conduction of heat in dielectric liquids and has obtained satisfactory results.

The present authors (Iida, Guthrie, and Morita 1982) have reconsidered the model theory of Andrade and have proposed an expression for the viscosity of liquid metals in terms of more fundamental physical parameters such as the pair distribution function and the average interatomic frequency[†].

In the derivation of the viscosity equation by Andrade, the following problems were re-examined by the authors: (a) his assumption of quasi-crystalline structures for liquids; (b) his neglect of atomic migration or self-diffusion; (c) his employment of Lindemann's formula for the frequency of interatomic vibration of liquids; (d) his derivation of the numerical factor 4/3.

[†] This idea originated with Takeuchi and Iida (Iida 1970; Takeuchi and Misawa 1971; Takeuchi and Iida 1972).

Our treatment for these controversial problems is as follows. (a) The structure of liquids, i.e. the distribution of atoms in liquid metal, may be represented by the pair distribution function $g(r)$ obtained experimentally. (b) The distribution of atoms represented by $g(r)$ provides only the time-averaged distribution of atoms. On a microscopic time scale, atoms in the liquid state can easily diffuse due to a fluctuation in temperature or kinetic energy. In the authors' approach, however, the phenomenon of free movement or diffusion of atoms is not directly considered, but an assumption is introduced that the frequency of atomic vibrations decreases with increasing temperature (this is a time-averaged, or apparent, atomic frequency). (c) Lindemann's formula, as modified by Iida et al., is used for the frequency of vibration of liquids. (d) Osida's treatment for the thermal conductivity of dielectric liquids is applied to the deduction of a numerical factor.

The phenomenon of viscosity in monatomic liquids, in the author's work, as in Andrade's, is based on the transfer of momentum of atomic vibrations from one layer to a neighbouring one. In other words, momentum transfer is produced by collisions or temporary unions with neighbours at every displacement of atomic vibrational motion. Momentum is therefore transported in the presence of a velocity gradient in the liquid. As such, our treatment of the mechanism of viscosity is similar to that of Andrade. We can also introduce this modelling approach to readily formulate the viscosity of monatomic liquids. Further, to obtain a tractable solution to an evaluation of these momentum interchanges, it is assumed that all atoms in a state of vibration are identical harmonic oscillators. While this assumption is only approximate, it is supported by the fact that both solid and liquid metals on either side of their melting points exhibit molar heat capacity values similar to those of harmonic oscillators (i.e. $C_V = 3R$). Metals also have similar phonon dispersion curves based on inelastic scattering data of neutrons (e.g. Egelstaff 1967d; Larsson, Dahlborg, and Jovic 1965; Larsson 1968; Cocking 1969).

We now imagine that a monatomic liquid flows in the x-axis direction with a velocity v and a velocity gradient dv/dz, i.e. the flow rate increases by an amount dv for each increment of distance dz in the z-axis direction perpendicular to the x–y plane. When an atom (A) at a given position a distance l away from the surface $z = z_0$ (the x–y plane) of equal velocity vibrates at an inclination θ to the z-axis, the atom (A) makes contact, in the case of a liquid of close-packed structure, at every vibration with two atoms in neighbouring planes which are at distances r away from the atom (A), where $l \le r$. Referring to Fig. 6.21, the atom (A) conveying momentum crosses the plane $z = z_0$ twice for each complete vibration. Thus, in the case of close-packed liquids, the transfer of momentum per vibration per atom is given (Andrade 1934; Gotō, Hirai, and Hanai 1964) by

$$4m\left(\frac{dv}{dz}\right)r\cos\theta \qquad (6.25)$$

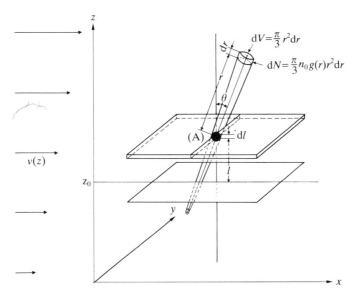

FIG. 6.21. Explanation of viscosity as communication of momentum at every displacement of an atom (A) (after Iida, Guthrie, and Morita 1982).

where m is the mass of the atom. Since there are 12 nearest neighbours in a close-packed structure of liquid metals, the solid angle occupied by one atom (refer to the atom (A)) can be taken to be equal to $\pi/3$ ($= 4\pi/12$). The number of atoms dN within a spherical shell of solid angle $\pi/3$ between the distances r and $r + dr$ from the reference atom (A) is given by

$$\frac{\pi}{3} n_0 g(r) r^2 \, dr \qquad (6.26)$$

where n_0 represents the average number density and $g(r)$ the pair distribution function. Consequently, the transfer of momentum per vibration per atom is given by the product eqns (6.25) and (6.26)

$$\frac{4\pi}{3} m \left(\frac{dv}{dz}\right) n_0 g(r) r^3 \, dr \cos \theta \qquad (6.27)$$

Since all directions of interatomic vibration may be regarded as equally probable, the fraction of atoms vibrating at angles between θ and $\theta + d\theta$ is $\sin \theta \, d\theta$ (i.e. $2 \times 2\pi r \sin \theta \, d\theta / 4\pi r^2$), where $\theta \leq \cos^{-1}(l/r)$. The number of atoms in a tube of unit cross section between $z = l$ and $z = l + dl$ planes is $n_0 dl$, so that the number of atoms vibrating at angles between θ and $\theta + d\theta$ within the reference tubular volume element is

$$n_0 \, dl \sin \theta \, d\theta \qquad (6.28)$$

Since viscous forces arise as a result of the rate of momentum interchange between atoms of adjacent layers, the frequency of atomic vibrations is needed. In our approach, we do not consider directly the phenomenon of movement of atoms by diffusion. We therefore compensate by taking a time-averaged, or apparent, frequency v of atomic vibrations. On melting, an increase in temperature fluctuations or kinetic energy fluctuations is to be expected in view of the increase in disorder and thermal motion in the liquid compared to the solid. Assuming that an atom vibrates around its mean position until a kinetic energy fluctuation causes it to diffuse in a linear manner to its next place of oscillation, the probability $W(\phi)$ of a thermal fluctuation ϕ occurring is given (e.g. Landau and Lifshitz 1958) by the Gaussian approximation,

$$W(\phi)\,d\phi = (2\pi\overline{\phi^2})^{-\frac{1}{2}}\exp\left(-\frac{\phi^2}{2\overline{\phi^2}}\right) \tag{6.29}$$

The fluctuation in kinetic energy ϕ can be replaced by

$$\phi = \frac{E - \overline{E}}{\overline{E}} \tag{6.30}$$

where $\overline{E}$ is the average kinetic energy, i.e. $\overline{E} = 3kT/2$, and E is the kinetic energy required for the thermal fluctuation. Supposing that E is the kinetic energy equivalent to the boiling point T_b, E is given by

$$E = \frac{3}{2}kT_b \tag{6.31}$$

Similarly, $\overline{\phi^2}$ can be represented by[†]

$$\overline{\phi^2} = \frac{\overline{E^2} - \overline{E}^2}{\overline{E}^2} \approx \frac{4}{3} \tag{6.32}$$

Using these relations, we therefore have

$$W(\phi)\,d(\phi) = \left(\frac{3}{8\pi}\right)^{\frac{1}{2}}\exp\left\{-\frac{3}{8}\left(\frac{T_b - T}{T}\right)^2\right\}d\left(\frac{T_b - T}{T}\right) \tag{6.33}$$

Consequently, the probability, denoted by $P(T)$, that the atom will stay in a state of oscillation around a fixed coordinate position is given by

$$P(T) = 1 - \int\left(\frac{3}{8\pi}\right)^{\frac{1}{2}}\exp\left\{-\frac{3}{8}\left(\frac{T_b - T}{T}\right)^2\right\}d\left(\frac{T_b - T}{T}\right) \tag{6.34}$$

[†] $\overline{E^2} - \overline{E}^2 = kT^2 c_V \approx 3k^2T^2$, where c_V is the heat capacity per atom at constant volume.

or

$$P(T) = 1 - \int_{\psi}^{\infty} (2\pi)^{-\frac{1}{2}} \exp\left(-\frac{\psi^2}{2}\right) d\psi \tag{6.35}$$

where

$$\psi = \frac{\sqrt{3}}{2}\left(\frac{T_b - T}{T}\right)$$

We see that the probability function $P(T)$ becomes less than unity with increasing temperature. Assuming that the time-averaged frequency of interatomic vibrations at equilibrium bulk temperature T is given by

$$v = v_0 P(T) \tag{6.36}$$

where v_0 is a constant (v_0 corresponds to the frequency of oscillation when no net displacement, i.e. diffusion, of the oscillating atom takes place), we are finally able to write an expression for the shear stress τ or viscous force per unit area. Thus, by multiplying eqns (6.27), (6.28), and (6.36) and integrating with respect to r, l and θ, we have

$$\tau = \frac{4\pi}{3} v_0\, P(T) mn_0^2 \left(\frac{dv}{dz}\right) \int_0^a g(r) r^3\, dr \int_{-r}^{r} dl \int_0^{\cos^{-1}\frac{l}{r}} \sin\theta \cos\theta\, d\theta$$

$$= \frac{8\pi}{9} v_0 P(T) mn_0^2 \left(\frac{dv}{dz}\right) \int_0^{\frac{a}{r}} g(r) r^4\, dr \tag{6.37}$$

where a is the distance over which the transfer of momentum takes place. Comparing eqn (6.37) with Newton's law of viscosity by eqn (6.1), we see that

$$\mu = \frac{8\pi}{9} v_0 P(T) mn_0^2 \int_0^{\infty} g(r) r^4\, dr \tag{6.38}$$

It can be assumed that momentum interactions occur mostly between nearest-neighbour atoms. We can estimate their maximum distance of separation as being represented by the minimum between the first and second peaks in the $g(r)$ curve of Fig. 6.22, and use this as the upper limit a of the integral in eqn (6.38). The integral itself can be estimated to a good approximation using the empirical relation

$$\int_0^{\frac{a}{r}} g(r) r^4\, dr \approx 1.6 g(r_m) r_m^4 (r_m - r_0) \tag{6.39}$$

where r_m and r_0 are the positions of the first peak and its left-hand edge in the

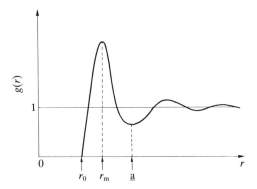

FIG. 6.22. Positions of the parameters, r_0, r_m, and $\underline{a}$ in a pair distribution function curve.

$g(r)$ curve, respectively, of the liquid metal under consideration (see Fig. 6.22). Substituting eqn (6.39) into eqn (6.38), we have the desired expression for the viscosity of liquid metals (in units of mPa s)

$$\mu \approx 4.5 v_0 P(T) m n_0^2 \ g(r_m) r_m^5 (1 - r_0/r_m) \tag{6.40}$$

Since r_m and n_0 are given by the relations $r_m \approx a$ and $n_0 \approx a^{-3}$ (where a is the average interatomic distance), eqn (6.40) can be simplified to

$$\mu \approx 4.5 g(r_m)(1 - r_0/r_m) \frac{v_0 P(T) m}{a} \tag{6.41}$$

The average frequency of vibration of liquid metals at their melting temperatures is given by Lindemann's formula as modified by the authors (i.e. eqn (5.25)). Using the relation by eqn (5.25), we have an expression for the melting-point viscosity μ_m

$$\mu_m \approx 4.5 \beta g(r_m)(1 - r_0/r_m) \frac{v_L m}{a} \tag{6.42}†$$

Since the average values of β, $g(r_m)$ and r_0/r_m at the melting points are 0.55, 2.64 and 0.82, respectively, eqn (6.42) can be roughly approximated by

$$\mu_m \approx 1.2 \frac{v_L m}{a} \tag{6.43}$$

This relation is then very similar to Andrade's formula by eqn (6.17) for melting-point viscosity of monatomic liquids.

Viscosity values calculated from eqn (6.38) depend upon the integral term appearing there. Since its evaluation is rather complex and arbitrary, liquid metal viscosities were calculated using eqn (6.40). Calculated values for

† At the melting temperature (see subsection 5.4.(3))
$$v_0 P(T) = \beta v_L$$

melting-point viscosities of various liquid metals are given in Table 6.3. There they are compared with experimental data and those calculated from Andrade's formula. In Table 6.4, predicted temperature dependence of liquid viscosities for sodium, iron, and lead are compared with observed values. As is obvious from these tables, the agreement between calculation and experiment is very good. The improvement over Andrade's results is noticeable, especially for iron and gallium. However, since the evaluation of the integral in the $g(r)$ curve is somewhat imprecise, we estimate accuracies on predicted viscosities of about ± 10 per cent.

TABLE 6.3. *Comparison of the calculated and observed values for the viscosity of liquid metals at their melting points*

Metal	Viscosity (mPa s)			References
	$\mu_{cal}{}^a$	$\mu_{cal}{}^b$	μ_{obs}	
Na	0.68	0.62	0.70	(1)
Mg	1.22	1.39	1.25	(2)
Al	1.90	1.79	1.2–4.2	See Fig. 6.26
K	0.50	0.50	0.54	(1)
Fe	6.37	4.55	6.92	(3)
Co	5.93	4.76	4.1–5.3	See Fig. 6.28
Ni	5.64	4.76	4.5–6.4	See Fig. 6.27
Cu	4.07	4.20	4.34	(4)
Zn	2.65	2.63	3.50	(5)
Ga	2.00	1.63	1.94	(6)
Ag	3.53	4.07	4.28	(4)
In	1.97	1.97	1.80	(4)
Sn	2.04	2.11	1.81	(4)
Sb	2.23	2.68	1.43	(4)
Au	5.50	5.80	5.38	(7)
Hg	2.31	2.06	2.04	(8)
Tl	2.55	2.85	2.64	(9)
Pb	2.52	2.78	2.61	(4)
Bi	2.13	2.54	1.63	(4)

[a] Values calculated from eqn (6.40).
[b] Values calculated from Andrade formula:

$$\mu_m = 1.8 \times 10^{-4}\, \frac{(M T_m)^{\frac{1}{2}}}{V_m^{\frac{2}{3}}} \quad \text{(In mPa s)}$$

where M is atomic weight, T_m is absolute temperature of melting point; V_m is atomic volume at T_m.

References: (1) Andrade and Dobbs (1952); (2) Lihl *et al.* (1968); (3) Morita and Iida (1982); (4) Iida *et al.* (1975); (5) Iida, Satoh, Ishiura, Ishiguro, and Morita (1980); (6) Iida, Washio, Kumada, and Morita (1980); (7) Gebhardt and Wörwag (1951); (8) Iida *et al.* (1973); (9) Crawley (1968).

TABLE 6.4. *Comparison of the calculated and observed temperature dependence of viscosity for liquid sodium, iron, and lead.*

Metal	Temperature (K)	Viscosity (mPa s)		$P(T)$	$g(r_m)$
		μ_{cal}[a]	μ_{obs}		
Na	371	0.70	0.70	0.966	2.46[a]
	378	0.68	0.67	0.962	2.43
	473	0.53	0.45	0.893	2.19
Fe	1808	6.55	6.92	0.716	2.64[a]
	1833	6.25	6.64	0.709	2.56
	1873	5.68	6.22	0.699	2.46
	1923	5.26	5.76	0.687	2.35
Pb	601	2.69	2.61	0.980	3.10[b]
	613	2.66	2.50	0.977	3.08
	823	1.97	1.72	0.896	2.80
	1023	1.41	1.32	0.802	2.53
	1173	1.22	1.16	0.735	2.34

[a] Values calculated from eqn (6.40).
[b] Extrapolated values.
Sources of data: $g(r_m)$, Waseda (1980d); μ_{obs}(Na), Andrade and Dobbs (1952); μ_{obs}(Fe), Morita and Iida (1982); μ_{obs}(Pb), Iida *et al.* (1975).

Values for the parameters used in the calculation of viscosities are shown in Table 6.5. The probability function $P(T)$ was obtained from tabulated values of the Normal Distribution Curve. β, r_0 and r_m are given in Tables 2.1 and 4.6, respectively.

Djemili and co-workers (Djemili, Martin-Garin, Martin-Garin, and Hicter 1980; Djemili, Martin-Garin, Martin-Garin, and Desré 1981) have proposed an extension of the Takeuchi and Iida (1972) model to liquid binary alloys, in the form

$$\mu = \frac{8\pi}{9} v_0 n_0^2 \left[x_A^2 P(A)m_A \int_0^a g_{AA}(r)r^4 \, dr + x_B^2 P(B)m_B \int_0^b g_{BB}(r)r^4 \, dr \right.$$

$$\left. + x_A x_B \{ m_A P(A) + m_B P(B) \} \int_0^c g_{AB}(r)r^4 \, dr \right] \qquad (6.44)$$

where $P(A)$ and $P(B)$ are the proportions of vibrators A and B in the liquid, x_A and x_B the concentrations of components A and B, and a, b, and c the values of r

TABLE 6.5. *Values of the parameters used for calculating viscosities in Table 6.3.*

Metal	T_m (K)	T_b (K)	v_0 $(10^{12}\,s^{-1})$	$P(T)^a$	m $(10^{-26}\,kg\,mol^{-1})$	n_0^a $(10^{28}\,m^{-3})$	$g(r_m)^a$
Na	371	1151	1.99	0.966	3.82	2.43	2.46
Mg	923	1380	4.78	0.666	4.04	3.94	2.50
Al	933	2333	4.29	0.903	4.48	5.33	2.87
K	337	1035	1.10	0.963	6.49	1.28	2.36
Fe	1808	3003	5.56	0.716	9.27	7.58	2.64
Co	1765	3458	4.78	0.797	9.79	7.92	2.50
Ni	1728	3448	4.43	0.840	9.75	8.11	2.43
Cu	1356	2903	3.54	0.839	10.55	7.58	2.86
Zn	693	1203	3.16	0.738	10.86	6.06	2.50
Ga	303	2573	2.24	1.000	11.58	5.28	2.65
Ag	1234	2253	2.63	0.763	17.91	5.19	2.67
In	430	2373	1.48	1.000	19.07	3.69	2.67
Sn	505	2548	1.53	1.000	19.69	3.54	2.62
Sb	904	1913	1.58	0.833	20.21	3.20	2.35
Au	1336	2983	1.89	0.857	32.71	5.33	2.95
Hg	234	630	1.14	0.929	33.29	4.11	2.73
Tl	576	1730	1.06	0.959	33.92	3.35	2.82
Pb	601	2023	1.01	0.980	34.41	3.10	3.10
Bi	544	1833	0.91	0.980	34.70	2.90	2.57

[a] Values at the melting points; $g(r_m)$: extrapolated values. Data for $g(r_m)$ from Waseda (1980d).

corresponding respectively to the first minimum of the pair distribution functions $g_{AA}(r)$, $g_{BB}(r)$, and $g_{AB}(r)$.

In evaluating values for $P(A)$ and $P(B)$, Djemili *et al.* have taken the statistical thermodynamical representation of a simple liquid, proposed by Hicter *et al.* (Hicter, Durand, and Bonnier 1971), instead of choosing the Gaussian function of Takeuchi and Iida.

They have applied the above expression to the cadmium–indium system. The agreement of the viscosity values calculated using eqn (6.44) with experimental data is fairly good.

Incidentally, the factor of $(4/3)v$ in Takeuchi and Iida's expression (1972) should be amended to $(8/9)\pi v_0$ ($v_0 \approx 0.5\,v$).

6.5.2. Empirical equations

The temperature dependence of both viscosity and diffusivity data in liquid metals approximate an Arrhenius type relationship. For viscosity, this is

$$\mu = A \exp\left(\frac{H_\mu}{RT}\right) \tag{6.45}$$

where A and H_μ are constants. This relationship is shown for several liquid metals over a wide temperature range in Fig. 6.23.

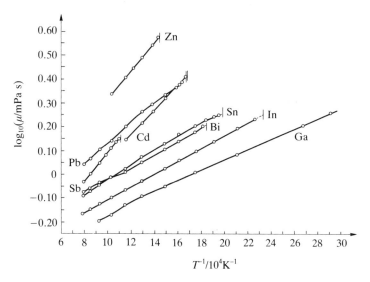

FIG. 6.23. Common logarithm of viscosity for liquid metals vs. reciprocal of the absolute temperature (after Iida, Morita, and Takeuchi 1975).

In order to estimate viscosity values for liquid metals, Grosse (1961b) has attempted to represent these constants (i.e. A and H_μ) using well-known parameters. According to Grosse's approach, the constant A can be obtained by using Andrade's formula for the melting-point viscosity. This constant then becomes

$$A = \frac{5.7 \times 10^{-2}\,(MT_{\mathrm{m}})^{\frac{1}{2}}}{V_{\mathrm{m}}^{\frac{2}{3}}\exp(H_\mu/RT_{\mathrm{m}})} \quad (\text{in cP})^{\dagger} \qquad (6.46)$$

Further, Grosse has indicated that a simple empirical relationship exists between H_μ for liquid metals and their melting points T_{m}, as illustrated in Fig. 6.24.

Iida et al. (1975) have also carried out a similar analysis and obtained the following relations for H_μ.

For normal metals: $\quad (H_\mu)_{\mathrm{n}} = 1.21\,T_{\mathrm{m}}^{1.2}$

For semi-metals: $\quad (H_\mu)_{\mathrm{s}} = 0.75\,T_{\mathrm{m}}^{1.2}$ $\qquad (6.47)$

Figure 6.25 shows these relationships. Incidentally, this classification may be related to differences in the structure factor between these two groups of liquid metals.

$\dagger$ $1\,\mathrm{cP} = 1\,\mathrm{mPa\cdot s}$

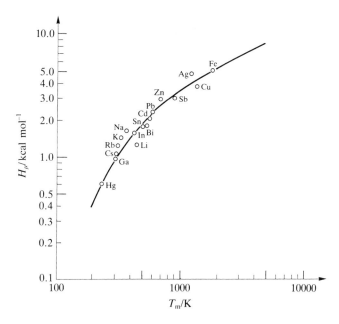

FIG. 6.24. Energy of activation for viscous flow vs. melting point (after Grosse 1961b).

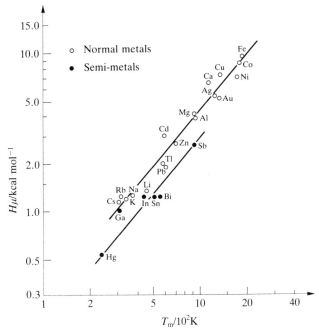

FIG. 6.25. Apparent activation energy viscous flow, H_μ, vs. melting point of liquid metal.
(after Iida et al. 1975)

6.6. EXPERIMENTAL DATA FOR THE VISCOSITIES OF PURE LIQUID METALS

It is very difficult to state definitely the accuracy of viscosity measurements for liquid metals. Errors of ± 1 to ± 20 per cent would seem to be a fair estimate with the exception of a few metals. There are not many well-established data for liquid metal viscosities. Viscosity data for pure liquid metals are listed in Table 6.3. Data for others currently available are given in Table 6.6.

As shown in Fig. 6.26, experimental viscosity values for liquid aluminium exhibit very large discrepancies, and the values calculated from eqn (6.40)

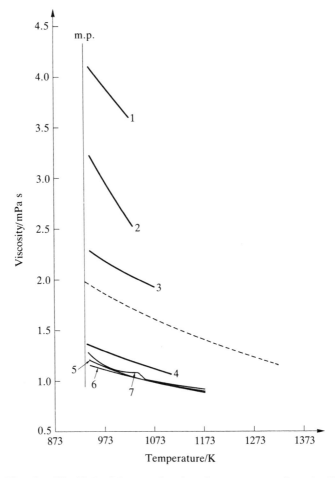

FIG. 6.26. Viscosity of liquid aluminium as a function of temperature, as determined by several workers: (1) Jones and Bartlett (1952–3); (2) Yao and Kondic (1952–3); (3) Yao (1956); (4) Rothwell (1961–2); (5) Gebhardt and Detering (1959); (6) Gebhardt, Becker, and Dorner (1954); (7) Lihl, Nachtigall, and Schwaiger (1968); – – – – predicted from eqn (6.40).

TABLE 6.6. *Viscosity of liquid alkali, lanthanide, and actinide metals at or near the melting point.*

Metal	Temperature (K)	Viscosity (mPa s)
Li	453.15	0.602
Rb	311.65	0.674
Cs	301.35	0.686
La	1193.15	2.65
Ce	1077.15	3.20
Pr	1208.15	2.85
Yb	1097.15	2.67
U	1406.15	6.53
Pu	913.15	5.93

Sources of data: alkali metals, Andrade and Dobbs (1952); lanthanide and actinide metals, Wittenberg and DeWitt (1973).

might therefore be recommended as being more appropriate. An experimental re-examination of the viscosities of some liquid metals, in particular, nickel and cobalt, are also called for. Computed and observed values for the viscosities of liquid nickel and cobalt are given in Figs. 6.27 and 6.28.

6.7. VISCOSITY OF LIQUID ALLOYS

6.7.1. Viscosity of dilute liquid alloys

In practical operations, the physico-chemical properties of dilute liquid alloys are important. A number of experimental studies have been carried out on the viscosities of dilute alloys, particularly on dilute iron-based alloys. Unfortunately, experimental results are frequently inconsistent. Considering an experimental error of the order of 20 per cent, the great majority of data for dilute liquid alloys lacks reliability.

Iida, Kasama, Morita, Okamoto, and Tokumoto (1973) have carefully measured the viscosities of mercury-based dilute binary alloys containing 1–3 at. % solute, using the capillary method. As shown in Fig. 6.29, the changes in the viscosity of pure liquid mercury are about 1–5 per cent with the addition of 1 at. % solute. Considering the experimental results of the amalgams and the main parameters which are assumed to dominate in determining the viscosity of a liquid alloy, differences in μ between the pure metal and a dilute alloy, especially at elevated temperatures, is comparatively small.

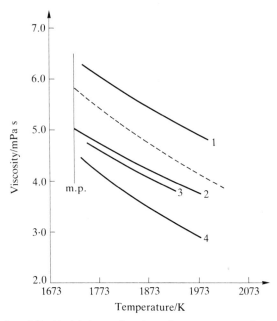

FIG. 6.27. Viscosity of liquid nickel as a function of temperature, as determined by several workers: (1) Vertman and Samarin (1960); (2) Cavalier (1963); (3) Schenck, Frohberg, and Hoffman (1963); (4) Samarin (1962); – – – – predicted from eqn (6.40).

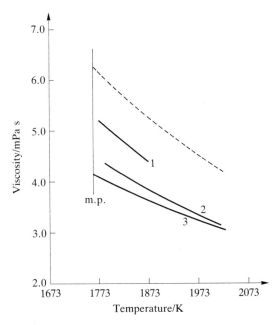

FIG. 6.28. Viscosity of liquid cobalt as a function of temperature, as determined by several workers: (1) Frohberg and Weber (1964); (2) Watanabe and Saito (1968); (3) Cavalier (1963); – – – – predicted from eqn (6.40).

Attempts have been made to calculate the viscosities of dilute liquid alloys from a theoretical standpoint (Okajima and Shimoji 1972; Shimoji 1977b). As yet, no theoretical formula provides reliable or precise data for the viscosities of various dilute liquid alloys because little is known of the interaction of dissimilar atoms.

An empirical relation for the viscosities of dilute liquid alloys is shown in Fig. 6.29. The parameter k_μ used in Fig. 6.29 is given by

$$k_\mu = \left(\frac{2a_M}{a_M + a_X}\right)^2 \left(1 + \frac{8.0 \times 10^6 |\Delta\chi|^2}{T_{m,x} T}\right) \tag{6.48}$$

where a is the average interatomic distance, χ is the electronegativity ($|\Delta\chi|$ = $|\chi_M - \chi_X|$), $T_{m,x}$ is the melting point, and the subscripts M and X refer to the base metal and solute elements, respectively. If the mechanism of viscosity is due to the communication of momentum between nearest neighbours, the lifetime of the oscillating atoms may be influenced by the differences in electronegativity of the base metal and the solute element, i.e the transfer of electrons between solvent and solute atoms.

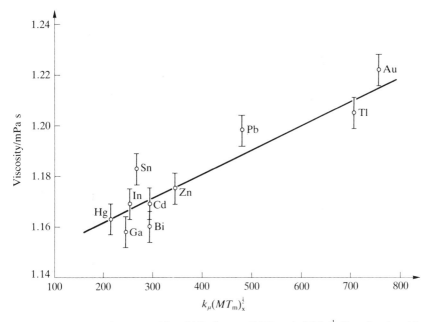

FIG. 6.29. Viscosities of mercury–1.0 at. % X alloys at 413 K vs. $k_\mu (MT_m)^{\frac{1}{2}}_x$. Data for viscosities from Iida et al. (1973).

6.7.2. Viscosity of binary liquid alloys

Numerous attempts have been made to describe the viscosities of binary liquid mixtures (e.g. Fort and Moore 1966; Moelwyn-Hughes 1961b). For example, Moelwyn-Hughes (1961b) proposed that

$$\mu_A = (x_1\mu_1 + x_2\mu_2)\left(1 - 2x_1x_2\frac{\Delta u}{kT}\right) \tag{6.49}$$

TABLE 6.7. *Relationship between excess viscosity of various binary liquid alloys and difference in the ionic radii (after Pauling)/integral enthalpy of mixing.*

Alloy	$\lvert d_1 - d_2\rvert$ (Å)	μ^E	H^E
Au–Sn	0.66	–	–
Ag–Sb	0.64	–	–
Ag–Sn	0.55	–	–,+ [a]
Al–Cu	0.46	–	–
Au–Cu	0.46	0, –	–
K–Na	0.38	–	+
Cd–Sb	0.35	–	+, – [a]
Cu–Sb	0.34	–	–
Ag–Cu	0.30	–	+
Hg–In	0.29	–,+ [a]	–
Cu–Sn	0.25	–	–
Al–Zn	0.24	–	+
Cd–Bi	0.23	–	+
K–Hg	0.23	+	–
Pb–Sb	0.22	–	+, – [a]
Mg–Pb	0.19	+, – [a]	–
Na–Hg	0.15	+	–
Al–Mg	0.15	+	–
Cd–Pb	0.13	–	+
Pb–Sn	0.13	0	+
Sb–Bi	0.12	–	+
Ag–Au	0.11	+	–
Fe–Ni	0.10	–,+ [a]	–
In–Bi	0.07	+	–
Mg–Sn	0.06	+	–
In–Pb	0.03	–	+
Sn–Zn	0.03	–	+
Sn–Bi	0.03	+	+

[a] The signs of the excess viscosity and the integral enthalpy of mixing change with composition (after Iida *et al.* 1976).

where μ_A is the viscosity of binary mixtures (alloys), x is the mole (atomic) fraction, Δu is the interchange energy[†] and the subscripts 1 and 2 refer to the components. Equation (6.49) may be written in the form of the excess viscosity[‡] μ^E

$$\mu^E = -2(x_1\mu_1 + x_2\mu_2)\frac{H^E}{RT} \tag{6.50}$$

This simple expression by Moelwyn-Hughes indicates that the signs (i.e. $+$, $-$) of the excess viscosity depend only upon those of the integral enthalpy of mixing. Table 6.7 shows the signs of the excess viscosity and the enthalpy of mixing. The excess viscosity μ^E appears to be related to the enthalpy of mixing

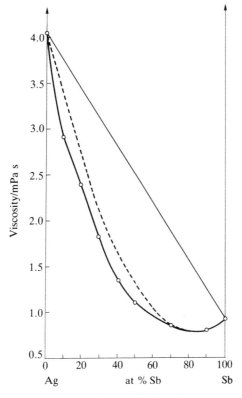

FIG. 6.30. Viscosities of silver–antimony alloys at 1273 K. ○ Experimental values (Iida); – – – – – values calculated from eqn (6.51) (after Iida, Ueda, and Morita 1976).

[†] $\Delta u = H^E/x_1 x_2 N_A$, where H^E is the (integral) enthalpy of mixing.
[‡] $\mu^E = \mu_A - (x_1\mu_1 + x_2\mu_2)$.

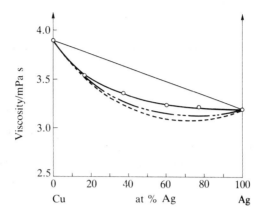

FIG. 6.31. Viscosities of copper–silver alloys at 1373 K. ○ Experimental values (Gebhardt and Worwag); ----- values calculated from eqn (6.51); ——— —— values calculated from eqn (6.52) (after Iida *et al.* 1976; Morita, Iida, and Ueda 1977).

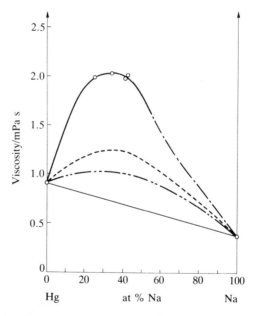

FIG. 6.32. Viscosities of mercury–sodium alloys at 643 K. ○ Experimental values (Degenkolde and Sauerwald);

——— — ———   values interpolated on the experimental data;
- - - - - -   values calculated from eqn (6.51);
——— —— ———   values calculated from eqn (6.52).
(after Iida *et al.* 1976).

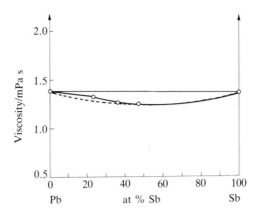

FIG. 6.33. Viscosities of lead–antimony alloys at 973 K. ○ Experimental values (Gebhardt and Köstlin); −−−−−− values calculated from eqn (6.51). (after Iida *et al.* 1976; Morita *et al.* 1977).

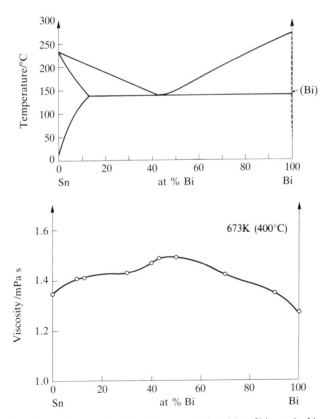

FIG. 6.34. Phase diagram (Hansen 1958) and isothermal viscosities of binary tin–bismuth system (after Iida 1970).

H^E, but the correlation of μ^E with H^E is not particularly satisfactory. Evidently, transport coefficients cannot be formulated in terms of only thermodynamic properties. Differences in atomic size and atomic mass of the components may also be important factors for transport properties. As indicated in Table 6.7, μ^E tends to become negative as the difference in atomic size increases.

Iida and co-workers (Iida, Ueda, and Morita 1976; Morita, Iida, and Ueda 1977) have proposed an expression for the excess viscosity of binary liquid alloys based on a phenomenological point of view. The equation is expressed in terms of some basic physical quantities, as follows

$$\mu^E = (x_1\mu_1 + x_2\mu_2)\left[-\frac{5x_1x_2(d_1-d_2)^2}{x_1d_1{}^2+x_2d_2{}^2} + 2\left\{\left(1+\frac{x_1x_2(m_1^{1/2}-m_2^{1/2})^2}{(x_1m_1^{1/2}+x_2m_2^{1/2})^2}\right)^{\frac{1}{2}}-1\right\} - \frac{0.12x_1x_2\Delta u}{kT}\right] \tag{6.51}$$

or

$$\mu^E = (x_1\mu_1 + x_2\mu_2)\left[-\frac{5x_1x_2(d_1-d_2)^2}{x_1d_1^2+x_2d_2^2} + 2\left\{\left(1+\frac{x_1x_2(m_1^{1/2}-m_2^{1/2})^2}{(x_1m_1^{1/2}+x_2m_2^{1/2})^2}\right)^{\frac{1}{2}}-1\right\} \right.$$
$$\left. -0.12(x_1\ln f_1 + x_2\ln f_2)\right] \qquad \text{(in cP)} \tag{6.52}$$

where d is the diameter of an atom (ionic radius after Pauling), m is the mass of an atom, and f is the activity coefficient. In the square brackets of the above equations, the first ($\mu^E_{h,d}$) and the second ($\mu^E_{h,m}$) terms represent the hard parts, and the third (μ^E_s) term the soft part of the friction constant for viscous movements of atoms. As is obvious from these equations, $\mu^E_{h,d}$ tends to become less negative as the difference between d_1 and d_2 increases. On the other hand, $\mu^E_{h,m}$ tends to become positive as the difference between m_1 and m_2 increases.

Values calculated from eqns (6.51) and (6.52) are shown in Figs. 6.30–6.33 together with experimental data. Calculated results coincide qualitatively with experimental data and, in particular, for regular or nearly regular solutions, excellent agreement between calculation and experiment is obtained.

Experimental data for the viscosity of binary liquid alloys are also in some confusion. For example, even in an amenable, low-melting alloy system such as lead–tin, the composition dependence of isothermal viscosity is inconsistent. In this system, some authors have reported anomalous viscosity changes at compositions corresponding to the eutectic and limits of solid solubility. On the other hand, other workers have obtained a linear or a slightly curved dependence on composition (Thresh and Crawley 1970).

It is sometimes pointed out that eutectic systems show negative deviations in

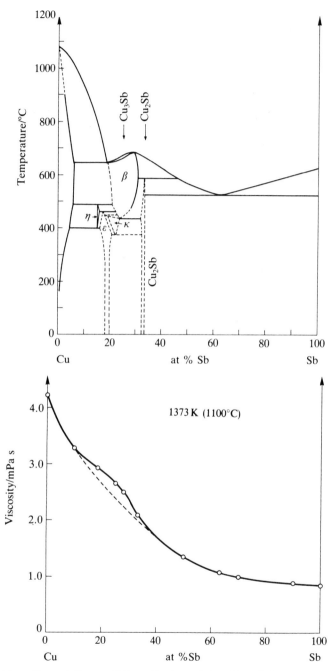

FIG. 6.35. Phase diagram (Hansen 1958) and isothermal viscosities of binary copper–antimony system (after Iida 1970).

isothermal viscosities. However, this relationship between excess viscosity and phase diagram has some exceptions. An example is shown in Fig. 6.34 together with its phase diagram. Compound-forming alloy systems are apt to show maxima in their isothermal viscosities in those composition ranges where compounds are formed in the solid state. This is indicated in Fig. 6.32. In addition, another example is shown in Fig. 6.35 together with its phase diagram. In the copper–antimony system, a maximum in the isothermal viscosity–composition curve is not conspicuous.

In general, discontinuous or sharp variations in the viscosities of metallic liquids with respect to temperature or composition should, at first, be doubted. Careful repetition of measurements is therefore necessary for any metallic liquids showing sharp changes in viscosities.

7

DIFFUSION

7.1. INTRODUCTION

Diffusion is the transport of mass from one region to another on an atomic scale. Diffusivities in the liquid state are much higher than diffusivities in the solid state. In the case of metals, diffusivities in the two states differ by a factor of 100 to 1000. The high atomic mobility of most metals just above their melting temperatures, with diffusivities of the order of $10^{-9}\,\mathrm{m^2\,s^{-1}}$, is one of the most characteristic properties of liquids.

A knowledge of diffusivities is needed for many fields of engineering. For example, for most metallurgical processes, heterogeneous chemical reactions play an important role. The rates of heterogeneous reactions, e.g. between two liquid phases such as slag and metal, are limited by the diffusion of the reactant species. Similarly, the distribution of solute elements during solidification also depends upon their diffusion motion. Although many other good examples are not hard to find, one can conclude, in general, that mass transfer through boundary layers is typical of heterogeneous reactions and that this mass transport is controlled by atomic diffusion phenomena.

Unfortunately, there are problems associated with experimental work relating to diffusion measurements in liquid metals. First, experimental data for the self-diffusivities of liquid metals are relatively scanty, mainly because of a lack of specific radio-isotopes. At present, it would appear that experimental self-diffusivity data are only available for twelve or thirteen liquid metals (see Tables 7.1 and 7.2). For a clear understanding of those phenomena related to diffusion, a study of self-diffusion in liquid metals is of critical importance. More reliable and accurate experimental data for the self-diffusivities of various liquid metals are therefore needed so as to test and develop theories, and to calculate solute or impurity diffusivities.

Remarkable progress has been made over the past fifteen to twenty years in the development of diffusion theories for liquid metals. The hard-sphere theory in particular seems to be able to provide an accurate means for calculating self-diffusivities in liquid metals. However, it may be premature to conclude its total success. A comparison between theory and experiment for various liquid metals is needed. Corresponding-states methods are presently believed to be more useful for estimating transport coefficients, although this does not coincide with the authors' views. For example, experimental values for liquid lead diffusivities (or liquid sodium viscosities) exhibit considerably larger deviations from self-diffusivity correlations (or from viscosity correlation) expected from the corresponding-states theory (see Figs. 6.20 and 7.5).

TABLE 7.1. *Comparison of experimental melting-point self-diffusivities with those calculated from eqns (7.18) and (7.26)*

Metal	Experimental $(10^{-9} m^2 s^{-1})$	Calculated $(10^{-9} m^2 s^{-1})$	
		eqn $(7.18)^a$	eqn (7.26)
Li	$6.19^{(1)}$, $6.80^{(2)}$, $5.76^{(3)}$, $6.80^{(3)}$, $5.98^{(4)}$	7.01	6.72
Na	$4.06^{(5)}$, $3.85^{(6)}$, $4.23^{(4)}$	4.24	4.10
K	$3.76^{(7)}$, $3.69^{(4)}$, $3.59^{(8)}$	3.85	3.71
Cu	$3.97^{(9)}$	3.40	3.18
Zn	$2.03^{(10)}$, $2.06^{(11)}$	2.55	2.45
Ga	$1.71^{(12)}$, $1.60^{(13)}$	1.73	1.64
Rb	$2.62^{(14)}$, $2.68^{(4)}$, $2.22^{(8)}$	2.68	2.58
Ag	$2.56^{(15)}$, $2.55^{(16)}$	2.77	2.68
Cd	$1.78^{(17)}$	2.00	1.94
In	$1.68^{(18)}$, $1.69^{(19)}$	1.77	1.72
Sn	$2.31^{(19)}$, $2.05^{(20)}$	1.96	1.86
Hg	$1.07^{(21)}$, $0.97^{(22)}$, $1.11^{(23)}$, $0.93^{(24)}$, $1.07^{(25)}$	1.07	0.93
Pb	$2.19^{(26)}$, $1.68^{(27)}$	1.67	1.60

[a] Calculated by Protopapas *et al.* (1973).
(1) Ott and Lodding (1965); (2) Murday and Cotts (1968); (3) Murday and Cotts (1971); (4) Larsson, Roxbergh, and Lodding (1972); (5) Meyer and Nachtrieb (1955); (6) Ozelton and Swalin (1968); (7) Rohlin and Lodding (1962); (8) Hsieh and Swalin (1974); (9) Henderson and Yang (1961); (10) Nachtrieb, Fraga, and Wahl (1963); (11) Lange, Pippel, and Bendel (1959); (12) Petit and Nachtrieb (1956); (13) Broome and Walls (1969); (14) Norden and Lodding (1967); (15) Yang, Kado, and Derge (1958); (16) Leak and Swalin (1964); (17) Ganovici and Ganovici (1970); (18) Lodding (1956); (19) Careri, Paoletti, and Vicentini (1958); (20) Onoprienko, Kuzmenko, and Kharkov (1966); (21) Hoffman (1952); (22) Nachtrieb and Petit (1956); (23) Meyer (1961); (24) Brown and Tuck (1964); (25) Broome and Walls (1968); (26) Rothman and Hall (1956); (27) Davis (1966).

In addition, calculated values for liquid iron viscosities using the viscosity correlation are approximately 30 per cent lower than reliable experimental data. To the authors' knowledge, experimental data for the self-diffusivities of pure liquid magnesium, aluminium, iron, cobalt, nickel, etc. have not yet been published. Experimental self-diffusivity data for these metals are badly needed.

A second difficulty with diffusion data is that while many experimental investigations have been carried out on solute diffusivities in liquid metals, the data are scarcely of sufficient accuracy. Similarly, while a variety of expressions describing solute diffusivities has been proposed, most of these equations are, unfortunately, unreliable in predicting solute diffusivities in many important liquid metal systems. Also, for some equations, the judgement of their validity is difficult because of extremely large errors in experimental data. This is especially true of diffusivity data for gases in liquid metals. The failure of theoretical calculations may well be due to a lack of knowledge of the physical state of solutes in liquid metals, e.g. their atomic size, the viscous forces or

TABLE 7.2. *Self-diffusivity data for pure liquid metals*

Metal	D_0 $(10^{-8}\,\mathrm{m^2\,s^{-1}})$	H_D $(10^3\,\mathrm{J\,mol^{-1}})$	Temp. range (K)	Ref.
Li	14.1	11.8		1
	9.4	9.62		2
	$D = \{5.76 + 0.036(T - T_\mathrm{m})\} \times 10^{-9}$			3
	$D = \{6.80 + 0.036(T - T_\mathrm{m})\} \times 10^{-9}$			3
	14.4	12.0	469–713	4
Na	11	10.2	372–500	5
	9.2	9.79		6
	8.6	9.29	376–557	4
K	17	10.7	340–490	7
	7.6	8.45	371–557	4
	$D = 5.344 \times 10^{-14} T^2 - 2.443 \times 10^{-9}$		354–868	8
Cu	14.6	40.6	1413–1533	9
Zn	8.2	21.3	723–873	10
	12	23.4	693–873	11
Ga	1.1	4.69	303–372	12
	$D = 6.01 \times 10^{-13} T^{\frac{3}{2}} - 1.60 \times 10^{-9}$		304–674	13
Rb	5.7	7.99	333–500	14
	6.6	8.28	330–510	4
	$D = 3.824 \times 10^{-14} T^2 - 1.479 \times 10^{-9}$		337–856	8
Ag	7.10	34.1	1275–1378	15
	5.8	32.0	1248–1623	16
In	2.89	10.2	448–1113	18
	3.34	10.7	500–901	19
Sn	3.02	10.8	540–956	19,20
Hg	1.3	4.85	276–364	21
	0.85	4.23	273–372	22
	$\log D = 1.854 \log T - 13.349$		273–567	23
	1.1	4.81	296–333	24
	$D = 4.34 \times 10^{-13} T^{\frac{3}{2}} - 4.81 \times 10^{-10}$		248–525	25
Pb	9.15	18.6	606–930	26

Reference numbers are the same as in Table 7.1. The review paper by Nachtrieb (1976) was also referred to.

friction constants between solute/solvent metal atoms, and atomic motion related to diffusion.

7.2. EXPERIMENTAL METHODS FOR MEASURING DIFFUSIVITIES

Diffusion occurs as a result of a concentration gradient or a density gradient. The flux of matter J, i.e. the amount of diffusing particles which passes per unit time through unit area of a plane perpendicular to the direction of diffusion, is generally proportional to a concentration gradient $\partial c/\partial x$ of the diffusing particles:

$$J = -D\frac{\partial c}{\partial x} \tag{7.1}$$

where c is the concentration, x is the distance in the direction in which diffusion occurs, and D is the diffusivity or diffusion coefficient (D is introduced as a proportionality factor). This equation is known as Fick's first law. The diffusing particles in binary systems, always flow in the direction of decreasing concentration gradient, and the negative sign is therefore introduced in order to make the diffusivity D positive. Diffusivity has the dimensions of (length)^2/(time), i.e. m^2/s in SI units.

On combining eqn (7.1) and the equation of continuity[†], we have

$$\frac{\partial c}{\partial t} = \frac{\partial}{\partial x}\left(D\frac{\partial c}{\partial x}\right) \tag{7.2}$$

If the diffusivity is independent of concentration, eqn (7.2) can be simplified to:

$$\frac{\partial c}{\partial t} = D\frac{\partial^2 c}{\partial x^2} \tag{7.3}$$

Equations (7.2) and (7.3) are known as Fick's second law.

Particular solutions to Fick's second law depend on the initial and the boundary conditions employed in experiments (see Bockris, White, and Mackenzie 1959; Edwards, Hucke, and Martin 1968; Nachtrieb 1972, 1976; Darken and Gurry 1953). These solutions provide information on concentration with respect to the time and system geometry.

The average value of the square of the linear displacement of particles after a time t is given by

$$\overline{x^2} = 2Dt \tag{7.4}$$

which is the diffusion law of Einstein and Smoluchowski (e.g. Moelwyn-Hughes, 1961c). This relation can be obtained by solving the diffusion equation, and also by considering a random walk process.

In the liquid state, the regular arrangement of atoms is destroyed, and atoms undergo Brownian motion according to the Langevin equation of motion. In this treatment, the diffusing atoms are impeded by the frictional forces of their

---

[†] The equation of continuity is given by

$$\frac{\partial c}{\partial t} = -\frac{\partial J}{\partial x}$$

This equation corresponds to the law of conservation of particles.

neighbours. The diffusivity can be formulated in terms of the friction coefficient ζ_f or the mobility U_D

$$D = \frac{kT}{\zeta_f} = U_D kT \qquad (7.5)$$

This is called the Einstein relation. According to the Einstein relation, a calculation of diffusivity devolves, in essence, to a calculation of friction coefficient ζ_f.

There are a number of methods for measuring liquid diffusivity, but the methods available for liquid metals are relatively few. The following methods have been used to measure liquid metal diffusivities: (a) capillary-reservoir, (b) diffusion couple, (c) shear-cell, (d) plane source, (e) electrochemical concentration-cell, (f) slow neutron scattering, and (g) nuclear magnetic resonance. Details of these methods have been reviewed (e.g. Bockris *et al.* 1959; Edwards *et al.* 1968; Nachtrieb 1972) and only outlines will be described here.

7.2.1. Capillary-reservoir method

In the capillary-reservoir method, the metallic sample containing the solute to be investigated is contained in a capillary tube of uniform diameter with one end sealed. This capillary is then immersed in a large bath, i.e. reservoir, of liquid solvent metal at a chosen temperature (Fig. 7.1). In the case of self-diffusivity measurements, the radio-isotope can be added to the liquid either in the capillary tube or in the bath. The capillary tubes used are usually a few millimetres or less in diameter, and from one to a few centimetres in length. After a measured time, the capillary is taken out of the reservoir, and the concentration of the specimen in the capillary is determined. The diffusivity can be calculated from this information and the length of the capillary, etc.

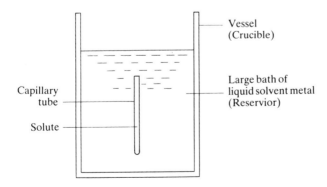

FIG. 7.1. The capillary-reservoir method.

(Grace and Derge 1955; Ma and Swalin 1960; Ejima, Inagaki, and Kameda 1966). In experiments to determine the diffusivity of liquid metals, convection currents are generally a critical problem. The capillary-reservoir technique largely allows one to avoid convection, thanks to the small diameter of the capillary. Another merit of this method is its relative insensitivity to slight vibrations or disturbances in the vicinity of the diffusion apparatus. Because of these advantages, the capillary-reservoir technique is the method most frequently adopted for measurements with liquid metals.

In order to obtain data with high accuracy, the use of a long capillary tube of fine diameter is desirable, even though attendant experimental difficulties increase.

7.2.2. Diffusion couple method

A capillary of uniform cross-section is half filled with solute or radioactive metal and half with solvent metal (Fig. 7.2). The lengths of the two specimens are much longer than the anticipated inter-diffusion distances in order that the system can be considered as being infinitely long on both sides of the interface. The specimens are rapidly heated to the desired temperature and the solute or tracer metal allowed to diffuse into the solvent metal. After a known time lapse, the capillary tube containing the specimens is cooled rapidly. Diffusivities can then be calculated from measurements of solute concentrations, or of tracer activities, at various points along the rods. A disadvantage of this technique is that the two specimens are in contact during the heating and cooling stages as well as during the diffusion period proper, so that some diffusion will occur during heating and cooling. The method is useful for cases where two metal samples melt at similar temperatures and where diffusivities are to be determined relatively close to their melting points. (see e.g. Calderon, Sano, and Matsushita 1971; Wanibe, Takagi, and Sakao 1975).

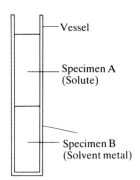

FIG. 7.2. The diffusion couple method.

7.2.3. Shear-cell method

So far, the shear-cell method has only been used for diffusivity measurements at low temperatures (e.g. mercury, gallium). The shear cell consists of two optically flat cylindrical disks mounted coaxially. Off-centre holes are drilled in these disks, which may be aligned to form the diffusion path. The solvent and solute or radioactive metal are kept separated until the run is begun by aligning the holes that make up the diffusion path. At the end of the run, the disks are again rotated to misalign the compartments. An advantage of this technique is its usefulness for diffusivity measurements at high pressure. The principal disadvantages of the method are that stirring can occur while the disks are being rotated at the start or end of an experiment, and that the equipment needed is relatively complex.

7.2.4. Plane source method

In this technique, one part of a diffusion cell comprises a thin planar section of about one millimetre or less (the 'plane source'), while the other is made up of a longer section of a few centimetres deep but much longer than diffusion distances (Fig. 7.3). At the start of a diffusion run, this thin piece disperses into the longer piece. The experimental procedure is similar to that of the diffusion couple method. (e.g. Morgan and Kitchener 1954).

7.2.5. Electrochemical concentration-cell method

Diffusivities may also be obtained from a liquid-metal concentration cell.
An advantage of the method is that it can substitute measurements of electrical quantities (i.e. current and voltage) for tedious chemical or radiochemical analyses. In other words, the technique allows direct measurements of

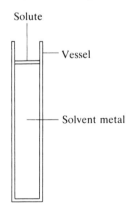

FIG. 7.3. The plane source method.

diffusivity in liquid metals, without solidifying the liquid metal specimen. A disadvantage of the technique is that only a restricted number of metal systems and concentration intervals are available for study. (e.g. Otsuka, Katayama, and Kozuka 1971; Kawakami and Goto 1976; Otsuka and Kozuka 1977).

Details of the electrochemical concentration-cell method are described by Edwards *et al.* (1968).

Diffusivity can also be determined using *the neutron scattering technique* and *the nuclear magnetic resonance (NMR) technique* (e.g. Nachtrieb 1972; Egelstaff 1967e). In the neutron scattering method, diffusivities are obtained by measuring, as a function of position and time, the intensity of emissions from a radioactive isotope of the element being investigated.

In the NMR method, diffusivities are determined by observing the diffusion rate for nuclear spins, i.e. the amplitude of the magnetization as a function of time (for a high diffusion rate, the amplitude decays rapidly).

In conclusion, while the neutron method uses measurements over distances of a few angstroms, the NMR method over distances of several hundred angstroms, and the tracer method over distances of millimetres, all these methods give, within experimental error, equivalent diffusivities for liquid metals.

7.3. THEORETICAL EXPRESSIONS FOR SELF-DIFFUSIVITY (DYNAMIC THEORIES OF LIQUIDS)

In atomic (or molecular) dynamic theories, atoms within a fluid are assumed to interact in accordance with a certain type of interatomic pair potential. Equations for dynamical properties are solved numerically using a computer of large storage capacity. These calculations provide useful information on structures and atomic motions in liquids.

7.3.1. Linear trajectory theory

In this theoretical analysis, the pair potential and friction coefficient are conveniently divided into two and three parts, respectively. The pair potential $\phi(r)$ is written (Rice and Nachtrieb 1967) in the form

$$\phi(r) = \phi_H(r) + \phi_S(r) \tag{7.6}$$

$$\left.\begin{aligned} \phi_H(r) &= \infty, \, r < \sigma \\ \phi_H(r) &= 0, \, r \geq \sigma \\ \phi_S(r) &= 0, \, r < \sigma \\ \phi_S(r) &= f(r), \, r \geq \sigma \end{aligned}\right\} \tag{7.7}$$

where $\phi_H(r)$ is the hard-sphere potential, $\phi_S(r)$ is the soft potential and σ is the effective core diameter. Similarly, the friction coefficient is separated into three component parts:

$$\zeta_f = \zeta_H + \zeta_S + \zeta_{SH} \tag{7.8}$$

where ζ_H, ζ_S and ζ_{SH} are friction coefficients, due to the hard-core collisions, to soft interaction between neighbouring atoms and to cross effect between hard and soft forces in the pair potential, respectively. According to Helfand (1961) and Davis and Palyvos (1967), the friction coefficients are given by

$$\zeta_H = \frac{8}{3} n_0 \sigma^2 g(\sigma)(\pi m k\,T)^{\frac{1}{2}} \tag{7.9}$$

$$\zeta_S = -\frac{n_0}{12\pi^2}\left(\frac{\pi m}{kT}\right)^{\frac{1}{2}} \int_0^\infty Q^3 \tilde{\phi}_S(Q)\tilde{G}(Q)\mathrm{d}Q \tag{7.10}$$

$$\zeta_{SH} = -\frac{1}{3} n_0 g(\sigma)\left(\frac{m}{\pi kT}\right)^{\frac{1}{2}} \int_0^\infty \{Q\sigma\cos(Q\sigma) - \sin(Q\sigma)\}\,\tilde{\phi}_S(Q)\mathrm{d}Q \tag{7.11}$$

where $g(\sigma)$ is the value of the pair distribution function at contact (i.e. $r = \sigma$), $\tilde{\phi}_S(Q)$ is the Fourier transform of the long-range part of the potential, and $\tilde{G}(Q)$ is the Fourier transform of $\{g(r) - 1\}$, with $g(r)$ the pair distribution function. Thus, from the Einstein relation and these expressions, the self-diffusivities are obtained:

$$D = \frac{kT}{\zeta_H + \zeta_S + \zeta_{SH}} \tag{7.12}$$

7.3.2. Small-step diffusion theory

According to the small-step diffusion theory of Rice and Kirkwood (1959), the soft part of the friction coefficient is given by

$$\zeta_S = \left\{ \frac{4\pi m n_0}{3} \int_0^\infty \nabla_r^2 \phi_S(r)g(r)r^2 \mathrm{d}r \right\}^{\frac{1}{2}} \tag{7.13}$$

In this approach, the self-diffusivity is expressed in terms of ζ_H and ζ_S as follows

$$D = \frac{kT}{\zeta_H + \zeta_S} \tag{7.14}$$

Several investigators (see Waseda 1980f) have calculated self-diffusivities of pure liquid metals using the linear trajectory and the small-step diffusion theories. The results of their theoretical analyses provide good qualitative agreement with experiment. Calculated results indicate that the soft part of the effective interatomic potentials play a dominant role in determining the magnitude of self-diffusivities in liquid metals. However, the accuracy of the calculations, or rather, the uncertainty of the pair potentials used, need more thorough examination.

7.4. DIFFUSION EQUATIONS BASED ON THE HARD-SPHERE THEORY

Metallic atoms exhibit symmetrical surrounding force fields which are approximately equivalent to the presence of hard spheres. During the last fifteen or twenty years, considerable progress has been made in calculating self-diffusivities in liquid metals through development of the hard-sphere theory.

Several expressions (in units of $cm^2 s^{-1}$) for self-diffusivity in liquid metals have been proposed, which are based on hard-sphere models.

(1) An expression by Vadovic and Colver (1970):

$$D = 0.365r \left(\frac{\pi k T}{m} \right)^{\frac{1}{2}} \frac{\eta_m/\eta}{9.385(T_m\rho/T\rho_m) - 1} \tag{7.15}$$

$$r = \left\{ \frac{3}{4} \left(\frac{\eta_m M}{\pi \rho_m N_A} \right) \right\}^{\frac{1}{3}}$$

where r is the atomic radius (i.e. $r = \sigma/2$).

(2) An expression by Faber (1972b):

$$D = 4.9 \times 10^{-6} \left(\frac{T}{M} \right)^{\frac{1}{2}} V^{\frac{1}{3}} \frac{(1-\eta)^3}{\eta^{\frac{5}{3}}(1-\eta/2)} \tag{7.16}$$

With $\eta_m = 0.45$ at the melting temperature, eqn (7.16) becomes

$$D_m = 4.0 \times 10^{-6} \left(\frac{T_m}{M} \right)^{\frac{1}{2}} V_m^{\frac{1}{3}} \tag{7.17}$$

(3) An expression by Protopapas, Andersen and Parlee (1973):

$$D = \sigma C_{AW}(\eta) \left(\frac{\pi R T}{M} \right)^{\frac{1}{2}} \frac{(1-\eta)^3}{8\eta(2-\eta)} \tag{7.18}$$

In the approach of Protopapas et al., values of σ, η, and $C_{AW}(\eta)$ can be calculated using the following relations:

$$\sigma = 1.126\sigma_m\{1 - 0.112(T/T_m)^{\frac{1}{2}}\} \tag{7.19}$$

in which σ_m is the value of σ at the melting point. With η_m taken as 0.472:

$$\sigma_m = \{6(0.472)M/\pi\rho_m N_A\}^{\frac{1}{3}}$$

$$= 1.41(M/\pi\rho_m N_A)^{\frac{1}{3}}$$

$$\eta = \frac{0.472\rho\sigma^3}{\rho_m\sigma_m^3} \tag{7.20}$$

The correction factor $C_{AW}(\eta)$ is obtained from Fig. 7.4, once the value η is known.

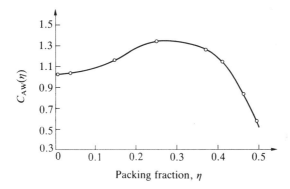

FIG. 7.4. The Alder–Wainwright correction C_{AW} as a function of the packing fraction η. The correction factor is defined as the ratio of the self-diffusion coefficient of a hard-sphere fluid to the value predicted by the Enskog theory (after Protopapas, Andersen, and Parlee 1973).

Equations (7.15), (7.16) and (7.18) generally give good results for the magnitudes and temperature dependence of the self-diffusivity for liquid metals. As an example, the calculated self-diffusivities for various liquid metals at their melting points using eqn (7.18) are compared with experimental values in Table 7.1. The computed results are in very good agreement with experimental data currently available. However, since experimental errors in diffusivities are of the order of ± 10 per cent, it is not particularly fruitful to make a detailed comparison of predicted and observed self-diffusivities.

7.5. EQUATIONS BASED ON PARTICULAR MODELS

7.5.1. Fluctuation model

Swalin (1959) has developed a theory for diffusion in liquids from the standpoint of fluctuation theory. According to Swalin, diffusion in liquids occurs by the movement of atoms through small, variable distances rather than by jumps of atoms into discrete-sized 'holes' of the order of the atomic diameter. These small, variable distances are caused by a local density fluctuation about the diffusing atoms. The energy required for atomic motion is the energy required for a density fluctuation and may be expressed in terms of a Morse function. The equation for self-diffusivity in liquid metals derived on this basis is represented either by

$$D = \frac{Zk^2T^2}{8hK_f} \qquad (7.21)$$

or

$$D = 1.29 \times 10^{-8} \frac{T^2}{\Delta_1^g H \alpha^2} \quad (\text{in cm}^2 \text{ s}^{-1}) \qquad (7.22)$$

where Z is the coordination number, h is Planck's constant, K_f is the force constant obtained from the data of Waser and Pauling (1950), $\Delta_1^g H$ is the enthalpy of vaporization, and α is related to the curvature of the potential vs. distance curve. Equations (7.21) and (7.22) suggest that the temperature dependence of the self-diffusivity is approximately proportional to the square of the absolute temperature, i.e. $D \propto T^2$.

Swalin warns that the fluctuation model may be more accurate in calculating relative values of the diffusivity than absolute values since, in calculating relative values, many of the quantities which are difficult to evaluate quantitatively, cancel out (Leak and Swalin 1964).

7.5.2. Significant structure theory

In the significant structure theory, the self-diffusivity is expressed (Breitling and Eyring 1972) as

$$D = \frac{kT}{\xi_A (V_S/N_A)^{\frac{1}{3}} \mu} \qquad (7.23)$$

in which V_S is the molar volume in the solid and ξ_A is a parameter expected to have values near six.

Breitling and Eyring (1972) have calculated self-diffusivities for six liquid metals and compared them with experimental data. Predicted values rarely differ from experimental values by more than an order of magnitude. Since the viscosities used were poor, their self-diffusivities would, as a result, also seem to be poor.

A number of other model theories for diffusion have been proposed. Other well-known examples are the activation-state theory or the jump diffusion model of Eyring and co-workers (Glasstone et al. 1941) the quasi-crystalline theory of Careri and Paoletti (1955), the free-volume theory of Cohen and

Turnbull (1959), the theory of Walls and Upthegrove (1964), the vibrational random-walk theory of Nachtrieb (1967) and so forth. These model theories are discussed in some excellent review papers (e.g. Edwards *et al.* 1968; Nachtrieb 1976) and will therefore not be repeated in this book.

In the case of self-diffusion, remarkable progress has been made in recent atomic dynamics calculations using the hard-sphere theory. This progress would appear to have somewhat reduced the role of model theories.

The objections to model theories are that, in general, they make drastic assumptions about the structure of a liquid and the mechanism of diffusion, and further, that they contain one or more *a posteriori* parameters which need to be fitted using experimental data.

7.6. SEMI-EMPIRICAL AND EMPIRICAL EQUATIONS

7.6.1. An equation based on the corresponding-states principle

Self-diffusivity data can also be correlated by use of the corresponding-states principle. According to Pasternak (1972), self-diffusivity can be expressed by

$$D = \frac{D^*}{(V^*)^{\frac{1}{3}}} \frac{(R\varepsilon/k)^{\frac{1}{2}}}{N_A^{1/3} M^{\frac{1}{2}}} V^{\frac{1}{3}} \tag{7.24}$$

where $\varepsilon/k = 5.2T_m$. The $D^*/(V^*)^{\frac{1}{3}}$ versus $1/T^*$ correlation is given in Fig. 7.5 (where $T^* = T/(\varepsilon/k)$) for the self-diffusivities of ten liquid metals. As can be seen, this correlation has a spread of ± 20 per cent for self-diffusivities if the high-temperature values for lead are neglected. The large deviations of lead from the correlation are, as yet, unexplained.

7.6.2. Empirical equations

In liquid metals, self-diffusivity increases exponentially with increasing temperature. The self-diffusivity may therefore be approximated by the expression

$$D = D_0 \exp\left(-\frac{H_D}{RT}\right) \tag{7.25}$$

where D_0 and H_D are constants. H_D is sometimes called the apparent activation energy.

Many workers (e.g. Larsson, Roxbergh, and Lodding 1972) have suggested that self-diffusivity in liquid metals at their melting points, D_m, may be expressed in terms of $(T_m/M)^{1/2} V_m^{1/3}$. Figure 7.6 indicates the relationship between experimental D_m data currently available and $(T_m/M)^{1/2} V_m^{1/3}$. From

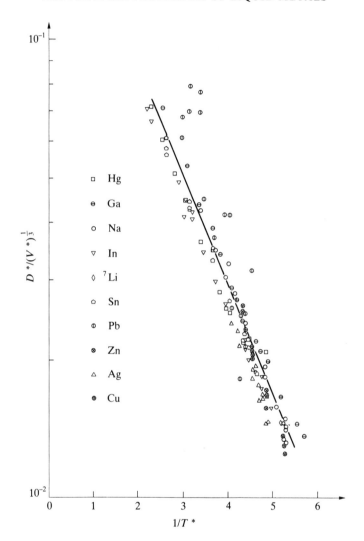

FIG. 7.5. Self-diffusivity correlation (after Pasternak 1972).

the results shown in Fig. 7.6, we can provide an empirical expression for self-diffusivities in liquid metals at their melting points:

$$D_m = 3.5 \times 10^{-6} \left(\frac{T_m}{M}\right)^{\frac{1}{2}} V_m^{\frac{1}{3}} \quad \text{(in cm}^2\text{/s)} \qquad (7.26)$$

In Table 7.1, values computed from equation (7.26) are presented and

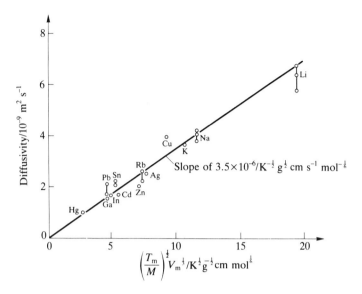

FIG. 7.6. Self-diffusivities in liquid metals at their melting points as a function of $(T_{\rm m}/M)^{1/2} V_{\rm m}^{1/3}$. Points linked by a vertical line represent two or three different experimental values for a single metal.

compared with experimental data. One finds that this empirical equation reproduces the data with good agreement.

The temperature dependence, i.e. the apparent activation energies H_D of self-diffusivity, may also be related to melting point temperatures. Figure 7.7 shows log H_D vs. $T_{\rm m}$ plots. From the empirical relationships between H_D and $T_{\rm m}$, values for H_D can be evaluated using:

$$\text{for normal metals: } (H_D)_{\rm n} = 2.50\, T_{\rm m}^{1.15}$$
$$\text{for semi-metals: } (H_D)_{\rm s} = 2.00\, T_{\rm m}^{1.15} \tag{7.27}$$

On combining eqns (7.25)–(7.27), we have an empirical equation for self-diffusivities in liquid metals:

$$D = \frac{3.5 \times 10^{-6}\, T_{\rm m}^{1/2}\, V_{\rm m}^{1/3}}{M^{1/2} \exp(-H_D/RT_{\rm m})} \exp\left(-\frac{H_D}{RT}\right) \tag{7.28}$$

7.7. RELATIONSHIP BETWEEN VISCOSITY AND DIFFUSIVITY

7.7.1. Sutherland–Einstein equation

Sutherland (1905) presented a correction to the Stokes–Einstein equation using hydrodynamic theory:

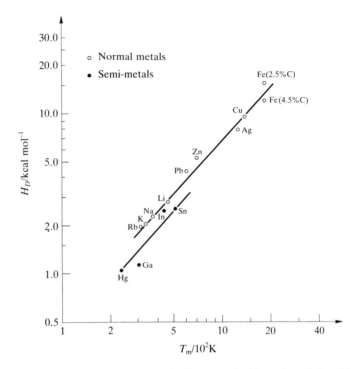

FIG. 7.7. Apparent activation energies for self-diffusivity in liquid metals vs. their melting points. Data for H_D from Nachtrieb (1967), and Larsson, Roxbergh, and Lodding (1972).

$$D = \frac{kT}{6\pi\mu r} \frac{(1 + 3\mu/\beta_s r)}{(1 + 2\mu/\beta_s r)} \tag{7.29}$$

where r is the radius of the diffusing particles and β_s is the coefficient of sliding friction between the diffusing particle and medium, which depends on the size of the diffusing particle. When the diffusing particle is large, compared with the particles of the medium, β_s equals infinity. When the radius of the diffusing particle is approximately equal to that of the medium, β_s becomes zero. In the two extreme cases, eqn (7.29) may be written

$$D = \frac{kT}{6\pi\mu r} \quad \text{when } \beta_s = \infty \tag{7.30}$$

$$D = \frac{kT}{4\pi\mu r} \quad \text{when } \beta_s = 0 \tag{7.31}$$

Equations (7.30) and (7.31) are generally known as the Stokes–Einstein formula and the Sutherland–Einstein formula, respectively. In these formulae, the friction coefficient is expressed in terms of viscosity and radius of the diffusing particle.

In order to answer the question of whether the diffusing species is in the form of an atom or ion in liquid metals, the Goldschmidt atomic radii and the Pauling ionic radii can be used for r. However, no conclusive answer has yet been presented to indicate which form is better.

7.7.2. Modified Stokes–Einstein equation

Several investigators (Larsson *et al.* 1972; Ott and Lodding 1965) have pointed out that the modified Stokes–Einstein formula is in excellent agreement with self-diffusivity data in liquid metals. The formula has the form

$$D = \frac{kT}{\xi(V/N_A)^{\frac{1}{3}}\mu} \tag{7.32}$$

in which ξ is a constant taking a value between 5 and 6 in cgs units.

The radius of a diffusing particle can be obtained on the basis of the hard-sphere model. From the definition of the packing fraction, we have

$$\left(\frac{V}{N_A}\right)^{\frac{1}{3}} = \left(\frac{4\pi}{3\eta}\right)^{\frac{1}{3}} r \tag{7.33}$$

With $\eta = 0.45$, eqn (7.33) becomes

$$\left(\frac{V}{N_A}\right)^{\frac{1}{3}} = 2.1r \tag{7.34}$$

Substituting this relation into eqn (7.32), we have

$$\frac{kT}{12.6\mu r} \lesssim D \lesssim \frac{kT}{10.5\mu r} \tag{7.35}$$

Equation (7.35) indicates that the modified Stokes–Einstein formula is equivalent to the Sutherland–Einstein formula by eqn (7.31), when the radius of a hard sphere is employed for the diffusing atom. Consequently, the Sutherland–Einstein equation using the hard-sphere radius will provide good results for self-diffusivity in liquid metals, even though it does not clarify the atomistic process of diffusion in liquids (Morita, Iida, and Ueda 1976).

By combining the modified Stokes–Einstein relation with Andrade's formula for melting-point viscosity, we have an expression for D_m in liquid metals:

$$D_m = \frac{kN_A^{1/3} \times 10^4}{5.7\xi}\left(\frac{T_m}{M}\right)^{\frac{1}{2}} V_m^{\frac{1}{3}} \tag{7.36}$$

Comparing eqn (7.36) with eqn (7.26), a ξ value of 5.8 for self-diffusivities in liquid metals is obtained.

7.8. EXPERIMENTAL DATA FOR SELF-DIFFUSIVITY IN PURE LIQUID METALS

Self-diffusivity data for pure liquid metals have been reviewed by Nachtrieb (1976) and Edwards *et al.* (1968). Experimental error in the data is estimated to be ± 10 per cent or more. Even in pure liquid mercury, discrepancies in self-diffusivity of about 10 per cent exist between the results of several investigators.

Experimental data for self-diffusivity in liquid metals, which seem to be currently available, are given in Table 7.2.

Predicted self-diffusivities by Protopapas *et al.* (1973) are listed in Table 7.3. No experimental data have been reported in the literature for these pure metals.

7.9. SOLUTE DIFFUSION IN LIQUID ALLOYS

Little theoretical work has been done on solute diffusion because knowledge of the fundamental quantities, i.e. the pair potentials and the pair distribution functions, of liquid alloys is insufficient.

Few investigators have calculated solute diffusivities in liquid alloys on the basis of hard-sphere models. Shimoji (1977*c*) has computed the ratios of solute to solvent diffusivities for dilute liquid alloys (metal–metal systems). Protopapas and Parlee (1976) have computed interdiffusivities for 12 gas–liquid metal systems. Roughly speaking, their calculated values appear to be reasonable with few exceptions. Unfortunately, a detailed comparison of the predicted and measured values of solute diffusivities in liquid alloys is extremely difficult, since the experimental errors are very large as a result of experimental difficulties. Some comparisons of their predictions with experimental results are given in Figs. 7.8–7.10.

7.9.1. Equations based on model theories

Solar and Guthrie (1972) have calculated hydrogen, oxygen, and carbon diffusivities in liquid iron using various model theories. In particular, the applicability of the small fluctuation models of Swalin and Reynik were tested in detail.

7.9.1.1. Swalin's model theory

Swalin's equation for solute diffusivity may be written in the form of the ratio of solute diffusivity D_i to solvent (or base metal) self-diffusivity $D_{S,M}$ (Swalin 1959; Leak and Swalin 1964; Swalin and Leak 1965).

TABLE 7.3. *Predicted self-diffusion coefficients for several elements which have not been investigated experimentally*

Metal	Temperature (°C)	ρ $(10^3\,\text{kg}\,\text{m}^{-3})$	D $(10^{-9}\,\text{m}^2\,\text{s}^{-1})$
Al	660.1[a]	2.37	4.87
	700	2.30	6.65
	800	2.33	7.09
	900	2.30	8.97
Ni	1455	7.859	3.90
	1500	7.76	4.61
	1550	7.65	5.35
	1600	7.54	5.96
Mg	650[a]	1.589	5.63
	700	1.570	6.78
	750	1.56	7.55
	800	1.55	8.73
	850	1.537	9.73
	900	1.525	11.1
	950	1.512	12.2
Sb	630.5[a]	6.50	2.66
	700	6.45	3.26
	750	6.415	3.76
	800	6.38	4.17
	850	6.35	4.60
	900	6.32	5.05
	1000	6.28	6.21
Cs	29.7[a]	1.84	2.31
	100	1.80	3.94
	200	1.737	6.78
	300	1.681	9.76
	500	1.562	17.2
	700	1.450	25.9
	800	1.400	30.9
Fe	1535[a]	7.103	4.16
	1600	7.038	4.91
	1700	6.938	5.85
	1800	6.838	6.90
	1900	6.738	8.08
	2000	6.638	9.46
	2100	6.538	10.5

[a] Melting temperatures.
(after Protopapas *et al.* 1973).

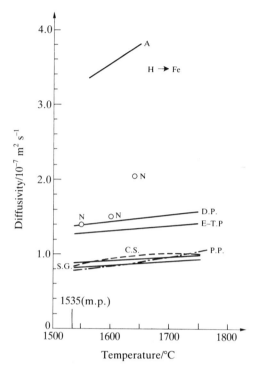

F<small>IG</small>. 7.8. Interdiffusion of hydrogen in liquid iron. A, Arkharov *et al.*; DP, Depuydt and Parlee; ETP, El-Tayeb and Parlee; SG, Solar and Guthrie; N, Nyquist; CS, calculation through the proposed theory implemented with Carnahan–Starling pair-correlation functions; PP, calculation through the proposed theory implemented with Protopapas–Parlee pair-correlation functions (after Protopapas and Parlee 1976).

$$\frac{D_i}{D_{S,M}} = \left\{ 1 - \frac{\varepsilon}{\lambda^2 K_f} \left(1 + \frac{2\lambda}{d} + \frac{2\lambda^2}{d^2} \right) \right\}^{-1} \qquad (7.37)$$

$$\varepsilon = \frac{\beta Z^E e^2}{d} \exp\left(-\frac{d}{\lambda} \right)$$

where λ is the screening radius (λ^{-1} is the screening constant), d is the distance between solute and solvent ions[†], Z^E is the relative solute to solvent valence, or the excess valence, β is a slowly varying quantity whose magnitude is a function of Z^E, and e is the electric charge.

In Solar and Guthrie's calculations, the following numerical values for the parameters of Swalin's equation were employed. The solutes were assumed to

---

[†] The distance between solvent ion centres has usually been used for d, because of a lack of knowledge of that between solute and solvent ions.

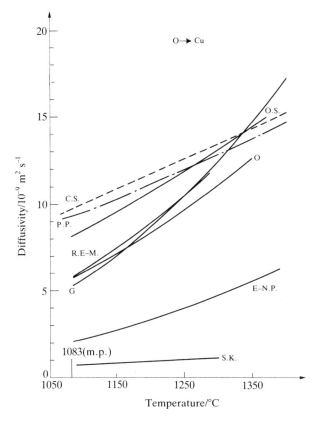

FIG. 7.9. Interdiffusion of oxygen in liquid copper. OS, Osterwald and Schwarzlose; RE–M, Rickert and El-Miligy; O, Oberg *et al.*; G, Gerlach *et al.*; E–NP, El-Naggar and Parlee; SK, Shurygin and Kryuk; CS, same as in Fig. 7.8; PP, same as in Fig. 7.8 (after Protopapas and Parlee 1976).

be in the form H^+, $O^=$ and C^{4+} (Bell and Lott 1963; *Handbook of Chemistry and Physics*, 1967–8), giving Z^E values of -1, -4, and $+2$ for hydrogen, oxygen, and carbon, respectively, where the charge of the iron particles is taken as $+2$. The corresponding β values were estimated as 1, 1.5, and 1 after Alfred and March (1955, 1956), and the screening radius for liquid iron calculated as 0.503 Å after Mott (1936). The liquid iron structural data used were those of Waseda, Suzuki, and Takeuchi (1969), i.e. $Z = 9.5$ and $d = 2.58$ Å. A force constant of 6×10^4 dyn cm^{-1} is extrapolated from the temperature dependence of K_f reported by Waser and Pauling (1950). In Table 7.4, the calculated values are listed together with experimental data. The agreement with the data is not satisfactory.

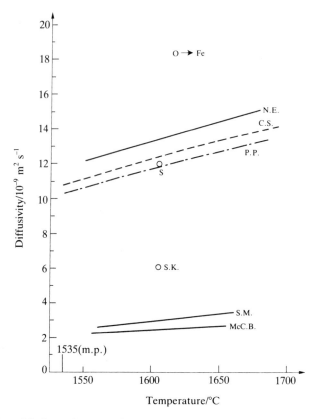

FIG. 7.10. Interdiffusion of oxygen in liquid iron. NE, Novokhatskii and Ershov; S, Schwerdtferger; SK, Shurygin and Kryuk; SM, Suzuki and Mori; McCB, McCarron and Belton; CS, same as in Fig. 7.8; PP, same as in Fig. 7.8 (after Protopapas and Parlee 1976).

7.9.1.2. Reynik's model theory (a semi-empirical small fluctuation theory)

According to Reynik (1969), the solute diffusivity is given by

$$D_i = 2.08 \times 10^9 \, Z x_0^2 T - 1.72 \times 10^{24} Z x_0^4 K_f \qquad (7.38)$$
$$x_0 = d - (r + r_i)$$

where r and r_i are the effective atomic radii of solvent and solute, respectively.

Solar and Guthrie (1972) have illustrated the solute diffusivities in liquid iron as a function of x_0. Figure 7.11 demonstrates this relationship for three liquids having force constants of 4×10^4, 5×10^4, and 6×10^4 dyn cm^{-1}, respectively (the other parameters being taken equal to the structural data of liquid iron). As seen, for low values of x_0, the D curves are asymptotic to the temperature term of eqn (7.38), and the magnitude of the force constants have

TABLE 7.4. *Diffusion coefficients of dissolved hydrogen, oxygen, and carbon in pure liquid iron*

Solute diffusion coefficients
(10^{-9} m^2 s^{-1})

Diffusing species	Theories calling (a) for ionic radii of diffusing particle								(b) for atomic radii	
	Stokes–Einstein	Sutherland	Li and Chang	Eyring		Walls and Upthegrove	Cohen–Turnbull	Swalin	Reynik	Expt. Data
				b.c.c. liq.	f.c.c. liq.					
Hydrogen	~ 300 000	~ 450 000	~ 720 000	~ 720 000	~ 480 000	~ 10^7	47.04	12.19	−44.31	92
Oxygen	2.31	3.46	5.44	5.44	3.62	3.43	15.34	9.79	+4.40	14
Carbon	19.05	28.58	44.88	44.88	29.92	50.84	22.72	14.33	+50.45	10

Data from Solar and Guthrie (1972) and Gourtsoyannis (1978).

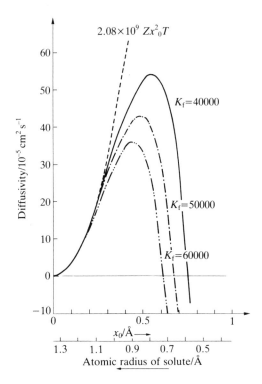

FIG. 7.11. Plot of predicted solute diffusion coefficients (10^{-5} cm^2 s^{-1}) in liquid iron vs. corresponding values of diffusion displacement x_0 for various values of force constant K_f appearing in eqn (7.38). Solute radii (Å) corresponding to these values of x_0 are also provided to illustrate the unsuitability of eqn (7.38) for small diffusing particles (after Solar and Guthrie 1972).

little influence on the predicted values. Most of the common ion's always fall in this range and it can be verified that the predicted diffusivities are usually within an acceptable range of measured values. However, the behaviour of the D curve at large x_0 values (i.e. smaller solutes) is startling: due to the increasing influence of the force constant term (proportional to $x_0{}^4 K_f$), the curve drops abruptly and even becomes negative. The elements oxygen, nitrogen, and hydrogen fall in this range. Thus the values predicted for hydrogen ($r_i = 0.46$ Å, $x_0 = 0.88$ Å) are all negative, which is, of course, incorrect. Even granted that the theory does not apply to hydrogen, the shape of the D curves at high x_0 values ($x_0 > 0.5$ Å), together with the uncertainty on atomic radii and lack of information on force constants, make useful predictions for other small solutes impossible.

Other impossibilities also invalidate the use of this semi-empirical small fluctuation theory. For example, the model does not allow for diffusion of

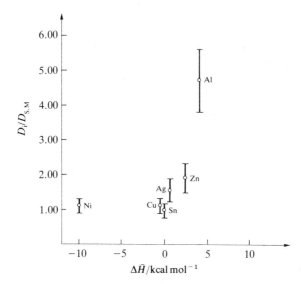

FIG. 7.12. $D_i/D_{S,M}$ vs. relative partial molar enthalpy of solutes in liquid tin (after Ma and Swalin 1960).

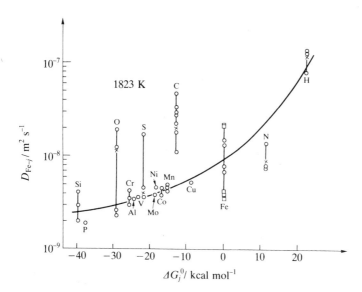

FIG. 7.13. Relation between diffusivity and standard free energy ΔG_i^0 of solution for various solute elements j in liquid iron at 1823 K (after Ono 1977). ○ observed values, □ predicted values, × recomended values of Bester and Lange.

solutes in liquid iron if their atomic radii are in excess of 1.34 Å $(d_{Fe} - r_{Fe})$
since x_0 then becomes negative[†].

One further discrepancy is the fact that Reynik's model (apparently
applicable to all liquids) predicts that atomic movements stop at temperatures
above 0 K, while all other approaches, including Swalin's, predict that such
movements stop at exactly 0 K in accord with common belief.

All model theories which have been tested by Solar and Guthrie (see Table
7.4) appear to be incapable of predicting liquid phase diffusivities for the
important systems carbon, oxygen, nitrogen, hydrogen, etc. in liquid iron or
sulphur, and oxygen in liquid nickel or copper. One must conclude that the
major discrepancies between various theoretical predictions and experimental
values for those systems may well be due to a lack of knowledge of the physical
state of solutes in liquid metals.

7.9.2. Empirical relations

Several empirical relations between solute diffusivity and thermodynamic
quantities for various elements in liquid metals have been presented (Suzuki

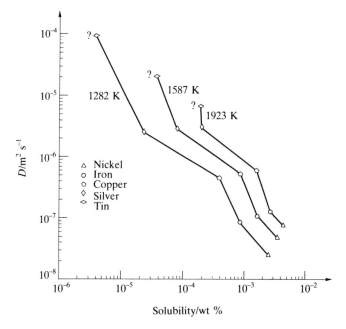

FIG. 7.14. Correlation between diffusion coefficients and hydrogen solubilities in liquid nickel,
iron, copper, silver, and tin (after Sacris and Parlee 1970).

[†] The atomic radii used were 1.24, 0.46, 0.60, and 0.77 Å for iron, hydrogen, oxygen, and carbon,
respectively.

and Mori 1971; Sacris and Parlee 1970; Kunze 1973; Depuydt and Parlee 1972; Ono 1977). Examples are given in Figs. 7.12–7.14.

Available data for solute diffusion or interdiffusion in binary liquid metals are summarized in review papers by Wilson (1965f), Edwards et $al.$ (1968), Ono (1977) (interdiffusivities in liquid iron) and Protopapas and Parlee (1976) (interdiffusivities of gases in liquid metals).

8

ELECTRICAL AND THERMAL CONDUCTIVITY

8.1. INTRODUCTION

In this final chapter, we review some of the more important electronic properties of liquid metals: their electrical conductivities or electrical resistivities, and their thermal conductivities. As is well known, solid metals are uniquely characterized by their high electrical and thermal conductivities. Similarly, liquid metals are also reasonably good conductors of electricity and heat. The high electrical and thermal conductivities of metals in their condensed states can be attributed almost entirely to freely moving electrons, or conduction electrons.

During the last three decades, theoretical considerations of electronic transport properties of liquid metals, and particularly their electrical conductivities, have made great progress thanks to the advent of pseudopotential theory. Notwithstanding this, only a few theoretical studies have been made on the subject of liquid metal thermal conductivities. In order to understand clearly the behaviour of electrons, which are the carriers of electric and heat currents in metals, some knowledge of quantum mechanics and quantum statistical mechanics is needed.

From a metallurgical point of view, relatively few studies have been carried out on the electrical properties of liquid metals and alloys. However, electrical conductivities of metallic liquids are of obvious importance to many liquid metal processing operations, e.g. electric furnace steelmaking and refining operations, electromagnetic stirring for melt cleanliness and microstructural control, and the electrowinning of aluminium from alumina in the Hall–Héroult cell. Thermal conductivities are also of significance in liquid metal processing operations. For instance, the generation of thermal natural convection phenomena in baths or furnaces can have important technological implications for the cast structure of that metal. Similarly, the melting rates of furnace or ladle additions, or the formation of protective thermal accretions around submerged nozzles, are all linked, in a fundamental sense, to the value of a metal's thermal conductivity. As a final example, the high thermal conductivities of liquid metals make them good media for removing heat from nuclear reactors.

Unfortunately, experimental investigations dealing with the thermal conductivities of liquid metals and alloys are far from satisfactory, and this is largely due to the extreme difficulty of making measurements in liquid metals in the absence of any convection or wetting problems, etc.

8.2. ELECTRICAL CONDUCTIVITY (ELECTRICAL RESISTIVITY) OF LIQUID METALS AND ALLOYS

8.2.1. Methods of electrical resistivity measurement

The electrical conductivity σ_e of a material is a unit of measurement representing its ability to carry electrical current in the presence of an applied voltage. It is expressed in terms of the current density ($A\,m^{-2}$) per unit electric field ($V\,m^{-1}$), its SI units of measurement being the mho or siemen per metre ($A\,V^{-1}\,m^{-1} = S\,m^{-1}$). Electrical conductivity σ_e is the reciprocal of Electrical resistivity ρ_e (Ωm), i.e. $\sigma_e = 1/\rho_e$.

The d.c. four-probe method and the rotating magnetic field method have been employed for measuring electrical resistivities of metallic liquids.

8.2.1.1. The d.c. four-probe method (direct method)

This technique is based on Ohm's law. Resistivity measurements are carried out by determining the potential drop across the liquid in a capillary tube of a known length and cross-section, while keeping a constant current density. In general, a cell constant of a given capillary is determined experimentally using a liquid of known electrical resistivity (usually mercury). A schematic diagram of resistivity cells is shown in Fig. 8.1. The principle and construction of the

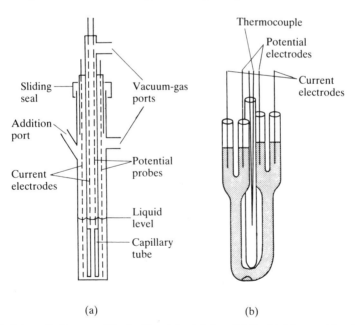

Sliding seal

Addition port

Current electrodes

Vacuum-gas ports

Potential probes

Liquid level

Capillary tube

Thermocouple

Potential electrodes

Current electrodes

(a)   (b)

FIG. 8.1. Schematic diagrams of the liquid-metal resistivity cell (after (a) Adams and Leach 1967; (b) Mera, Kita, and Adachi 1972).

method are both quite simple. As a result, the d.c. four-probe (potentiometric) technique has been most widely used for measuring electrical resistivities of liquid metals and alloys.

A serious problem with this technique is one of materials for the cell or capillary tube, and especially the electrodes. Although molybdenum, tungsten, or graphite can be used for electrodes, dissolution of electrodes in the liquid metal sample, or chemical reaction between the two, may take place. In order to eliminate these problems, a method (an improved four-probe method) which consists of four solid electrodes made of identical material to the liquid metal sample, and providing for a temperature gradient in the lower part of the cell as indicated in Fig. 8.2, has been developed (Ershov, Kasatkin, and Gavrilin 1976; Kita, Oguchi, and Morita 1978). However, the improved four-probe method is limited to pure liquid metals and alloys having a narrow liquid–solid range.

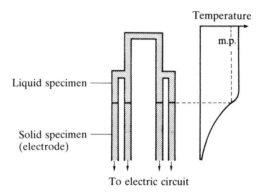

FIG. 8.2. Schematic diagram of cell and specimen of improved four-probe method (after Kita, Oguchi, and Morita 1978).

In resistivity measurements by these methods, degassing of the liquid metal specimen in the fine capillary tube is important.

Careful measurements using the four-probe method can provide accurate values of electrical resistivity of liquid metals.

8.2.1.2. *The rotating magnetic field method (indirect method)*

In measuring electrical resistivities of liquid metals and alloys at high temperatures, the rotating magnetic field method without electrodes has sometimes been used (Braunbek 1932; Roll, Felger, and Motz 1956; Roll and Motz 1957; Takeuchi and Endo 1962a, b; Ono and Yagi 1972; Ono, Hirayama, and Furukawa 1974; Samarin 1962).

This technique is based on the phenomenon that a cylindrical sample in a rotating magnetic field shows a torque depending linearly on the conductivity (i.e. the torque is inversely proportional to the electrical resistivity). A schematic diagram of the apparatus is shown in Fig. 8.3. The constant for the apparatus is determined experimentally by use of a Standard Sample (e.g. mercury).

Disadvantages are that the method requires density values of the liquid metal sample, and some corrections for calculating resistivities from experimental data.

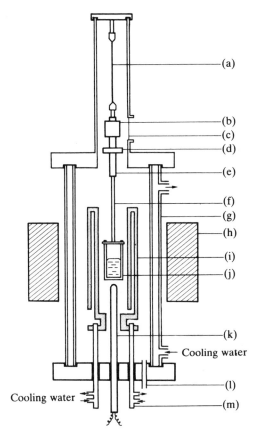

FIG. 8.3 Schematic cross-section of the apparatus for measuring electrical resistivity of liquid metals by the rotating magnetic field method (after Ono and Yagi 1972). ((a) Tungsten suspension wire; (b) Mirror; (c) Window; (d) Brass disk; (e) Brass rod; (f) Graphite rod; (g) Silica jacket; (h) Electromagnet; (i) Graphite heating element; (j) Alumina crucible with metal specimen; (k) Thermocouple (Pt. 30% Rh–Pt. 6% Rh); (l) Vacuum line; (m) Power supply)

8.2.2. Theoretical equations for electrical resistivity

The electrical resistivities of most metals in the liquid state just above their melting points are about 1.5–2.3 times as great as those of solid metals just below their melting points. This is caused by the fact that liquid metals with a relatively disordered arrangement of ions have higher resistivities (or lower conductivities) than crystalline solid metals with their more regular arrangement. Because the electron mean free path is shorter when the electrons are moving through the disordered liquids, these decreases in electrical conductivity are to be expected.

(1) Mott (1934) has proposed an equation for the ratio of liquid/solid conductivity $\sigma_{e,1}/\sigma_{e,s}$ at a metal's melting point, which is based on the simple assumption that the atoms in a liquid metal vibrate about slowly-varying mean positions with a frequency v_1. The equation is expressed in the form

$$\frac{\sigma_{e,1}}{\sigma_{e,s}} = \left(\frac{v_1}{v_s}\right)^2 = \exp\left(-\frac{80\Delta_s^1 H_m}{T_m}\right) \tag{8.1}$$

where $\Delta_s^1 H_m$ is the enthalpy of melting or the latent heat in kJ mol^{-1}, and v_s is the frequency of atomic vibration in the solid.

In spite of his simple treatment, the agreement obtained between theory and experiment for normal metals is surprisingly good.

Ziman (1961; see Faber 1972 c) has provided a reasonable interpretation for the ratio $\rho_{e,1}/\rho_{e,s}$ ($= \sigma_{e,s}/\sigma_{e,1}$), which is based on the modern theory of electron transport.

(2) According to the nearly free-electron model (the NFE model), the electrical resistivity ρ_e is given (Ziman 1961, 1964) by

$$\rho_e = \frac{12\pi V}{e^2\hbar v_F^2} \int_0^1 |U(Q)|^2 S(Q) \left(\frac{Q}{2k_F}\right)^3 d\left(\frac{Q}{2k_F}\right) \tag{8.2}$$

where e is the electronic charge, v_F and k_F are the Fermi velocity and Fermi wave vector, respectively, V is the atomic volume, $U(Q)$ is the pseudopotential, and $S(Q)$ is the (static) structure factor. Equation (8.2) is known as Ziman's formula.

Evans, Greenwood, and Lloyd (1971) and Evans, Gyorffy, Szabo, and Ziman (1973) have extended Ziman's formula to make it suitable for strong-scattering liquids.

$$\rho_e = \frac{12\pi V}{e^2\hbar v_F^2} \int_0^1 |t|^2 S(Q) \left(\frac{Q}{2k_F}\right)^3 d\left(\frac{Q}{2k_F}\right) \tag{8.3}$$

The t matrix can be expressed in terms of the phase shifts η_l of the various partial waves, as

$$t(q) = -\frac{2\pi\hbar^3}{m(2mE_F)^{\frac{1}{2}}V} \sum_l (2l+1) \sin \eta_l \exp(i\eta_l) P_l(\cos\theta)$$

where $q = |k'_F - k_F|$, E_F is the Fermi energy, and P is the Legendre polynomial.

Resistivity calculated through the use of eqn (8.3) is in good agreement with experimental data (Waseda 1980g).

(3) Takeuchi and Endo (1962a, b) have presented an expression for the electrical resistivity of liquid metals and alloys which makes use of the fluctuation scattering theory. Subsequently, Takeuchi and Endo's approach was extended by Tomlinson and Lichter (1969).

8.2.3. Experimental data

Table 8.1 lists experimental data on the electrical resistivities of liquid metals at their melting points, together with those for solid metals, and their ratio $\rho_{e,1}/\rho_{e,s}$. The accuracy of these data is good. For example, electrical resistivity data for pure liquid iron as measured by a number of different investigators are summarized in Table 8.2. As can be seen from Table 8.1, germanium and silicon, which are semiconductors in the solid state, show a remarkable decrease of their resistivities on melting, as they become metallic liquids. Similarly, antimony, bismuth, and manganese also show a decrease in resistivities on melting. This phenomenon of increasing conductivity can be attributed to an increase in the number of conduction electrons on melting.

The electrical resistivity of liquid metals increases linearly with increasing temperature, except for cadmium and zinc, which both have negative temperature coefficients just above their melting points. According to Kita *et al.* (1978), temperature coefficients of the resistivity for liquid iron, cobalt, and nickel are as follows (ρ_e is in units of $\mu\Omega$ cm):

$$\left. \begin{array}{l} \text{Fe: } \rho_e = 0.0154T(^\circ\text{C}) + 112.3 \\ \text{Co: } \rho_e = 0.0192T(^\circ\text{C}) + 9.18 \\ \text{Ni: } \rho_e = 0.0116T(^\circ\text{C}) + 70.2 \end{array} \right\} \qquad (8.4)$$

In general, a liquid metal's electrical resistivity is raised when foreign atoms are introduced to the melt. In other words, a liquid alloy's electrical resistivity shows generally positive deviations from additivity of its components. Some examples for the resistivities of liquid alloys are provided in Figs. 8.4–8.7. However, in the case of a liquid alloy system which is composed of polyvalent components, the resistivity sometimes shows a negative deviation from additivity of component resistivities. As indicated in Fig. 8.8, the isothermal

TABLE 8.1. *Electrical resistivities of liquid and solid metals at their melting points* T_m

Element	T_m (K)	$\rho_{e,s}$ ($\mu\Omega$ cm)	$\rho_{e,l}^a$ ($\mu\Omega$ cm)	$\rho_{e,l}/\rho_{e,s}$	α ($\mu\Omega$ cm K^{-1})	β ($\mu\Omega$ cm)	Temp. range, T_m to: (K)
Li	452			Insufficient	data		
Na	370.9	6.598	9.573	1.451	0.038	−4.5	573
K	336.8	8.32	12.97	1.56	0.064	−8.5	573
Rb	312.1	13.7	22.0	1.60	0.086	−4.8	373
Cs	301.6	21.7	36.0	1.66	Insufficient data		
Cu	1356	10.3 (5)	21.1	2.04	0.0089	9.1	1473
		9.4 (5)	20.0	2.1	0.0102	6.2	1873
Ag	1233.6	8.2 (5)	17.2(5)	2.09	0.0090	6.2	1473
Au	1336	13.68	31.2(5)	2.28	0.014	12.5	1473
Be	1557	No data					
Mg	924	15.4	27.4	1.78	0.0050	22.9	1173
Ca	1123	No data					
Sr	1043	No data					
Ba	983	~82	~134	1.62	Insufficient data		
Zn	692.6	16.7	37.4	2.24	Not linear		
Cd	594	17.1	33.7	1.97	Not linear		
Hg	234.2	18.4 (par.)	90.96	4.94 (par.)	Not linear		
		24.3 (perp.)		3.74 (perp.)			
Al	933.3	10.9	24.2	2.20	0.0145	10.7	1473
Ga	303	~56 c axis		0.45 c			
		~18 a	25.8	1.46 a	0.0195	19.9	670
		~8 b		3.12 b			
In	427	15.2	33.1	2.18	0.0255	22.2	1273
Tl	576	35.5	73.1	2.06	0.0292	56.3	1073
Si	1713	~2400	~81	0.034	0.113	−113	1820
Ge	1232	900	60	0.067			
		1200	63	0.053	Insufficient data		
		1100	85	0.077			
Sn	505	22.8	48.0	2.10	0.0249	35.4	1473
Pb	600.5	49.0	95.0	1.94	0.0479	66.6	1273
Sb	903.6	183	113.5	0.61	0.0270	87.9	1273
Bi	544	290.8	130.2	0.45	0.0570	99.2	1273
Se	493.6	~10$^{3.5}$	~10^7	~1000	Not linear		
Te	703.1	~5500 (par.)	~500	0.091 (par.)	Not linear		
		11000 (perp.)		0.048 (perp.)			
Mn	1516	66	40	0.61	Insufficient data		
Fe	1808	127.5	138.6	1.05			
		122	110	0.9	0.033	50	1973
Ni	1726	65.4	85.0	1.3	0.0127	63	1973
Co	1768	97	102	1.05	0.0612	−6	1973

[a] $\rho_{e,l}$ (in $\mu\Omega$ cm) is given by $\alpha T + \beta$ from the melting point to the temperature in column 8. (after Cusack and Enderby 1960).

TABLE 8.2. *Resistivities of several liquid metals at their melting points measured by different investigators*

Metal	Investigator	$\rho_{e,l}$ $(\mu\Omega\,cm)$	$\rho_{e,l}/\rho_{e,s}$	$d\rho_{e,l}/dT$ $(\mu\Omega\,cm\,K^{-1})$
Sn	Roll and Motz	48.0	2.10	0.0250
	Kita *et al.*	47.42		0.02673
Cu	Roll and Motz	21.1	2.04	0.00886
	Kita *et al.*	21.50	2.03	0.00837
Fe	Powell	138.6	1.09	
	Arsentiev *et al.*	135.1	1.06	0.0389
	Ono and Yagi	137.8	1.05	0.0434
	Kita *et al.*	135.9	1.06	0.0154
Co	Regeli *et al.*	102	1.05	0.061
	Ono and Yagi	126.5	1.21	0.0384
	Kita *et al.*	120.5	1.14	0.0192
Ni	Regeli *et al.*	85.0	1.3	0.013
	Ono and Yagi	85.0	1.47	0.0280
	Kita *et al.*	87.0	1.40	0.0116

(after Kita, Oguchi, and Morita 1978).

electrical resistivity of liquid mercury–indium alloy system has a negative deviation from that based on additivity of resistivities.

8.3. THERMAL CONDUCTIVITY OF LIQUID METALS AND ALLOYS

8.3.1. Methods of thermal conductivity measurement

Thermal conductivity λ is defined in a similar manner to electrical conductivity: λ = heat flow conducted across unit area per second per unit temperature gradient. Units of measurement for λ n the SI system are watt meter^{-1} kelvin^{-1} ($W\,m^{-1}\,K^{-1}$).

A variety of methods exists for measuring thermal conductivities of liquids (e.g. Kingery 1959; Ziebland 1969). However, the longitudinal heat-flow method (or the axial heat-flow method) has been most frequently used for good conductors such as metals. An apparatus for the technique is shown schematically in Fig. 8.9. The liquid metal sample is contained in a cell (see Fig. 8.9(b)) for which, usually, length/breadth $\approx$ 10, through which heat flows in a vertical direction. Heat losses in the horizontal direction are prevented by thermal shields.

The measurement of thermal conductivity is primarily concerned with the difference in temperature as related to heat flow. Thermal conductivity can be

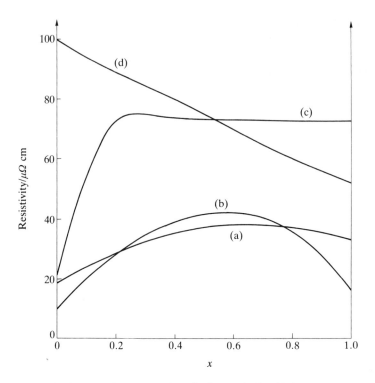

FIG. 8.4. Resistivity ρ_e vs. atomic concentration x for four typical liquid binary systems: (a) Ag–Au at 1473 K; (b) Na–K at 373 K; (c) Cu–Sn at 1473 K; (d) Pb–Sn at 673 K (after Faber and Ziman 1965).

computed from the measured heat flow and longitudinal temperature gradient. For example, Duggin (1969) has calculated the thermal conductivity of liquid mercury in a Pyrex cell[†] from the following equation:

$$\lambda = \frac{1}{A}\left(\frac{\dot{q}}{dT/dx} - \lambda_c A_c\right) \tag{8.5}$$

where $\dot{q}$ is the measured power input to the lower heater, λ_c is the thermal conductivity of the cell (Pyrex), A is the average cross-sectional area of the cavity in the cell containing the liquid metal, A_c is the average cross-sectional area of the (Pyrex) walls of the cell, and dT/dx is the longitudinal temperature gradient over the measured portion (B-E in Fig. 8.9(b)) of the cell.

[†] Pyrex cells were used because of ease of fabrication.

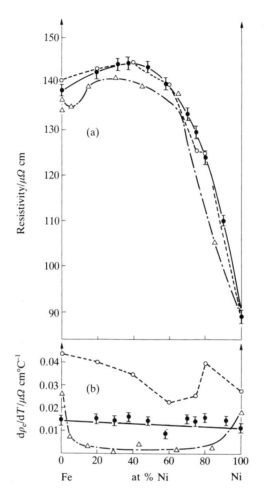

FIG. 8.5. (a) Electrical resistivity ρ_e at 1873 K, and (b) temperature coefficient $d\rho_e/dT$ for liquid iron–nickel alloys. ● Kita and Morita; ○ Ono and Yagi; △ Baum, Tyagunov, Gel'd, and Khasin (after Kita and Morita 1984).

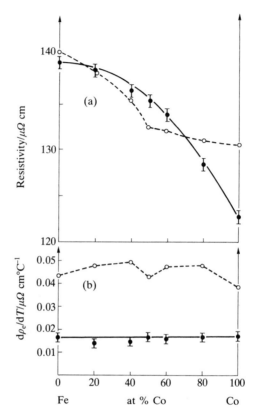

FIG. 8.6. (a) Electrical resistivity ρ_e at 1873 K, and (b) temperature coefficient $d\rho_e/dT$ for liquid iron–cobalt alloys. ● Kita and Morita; ○ Ono and Yagi; (after Kita and Morita 1984).

The major problem in thermal conductivity experiments is the measurement of heat flow. Similar measurements have been conducted for mercury, molten lead, and lead–tin alloys jacketed in stainless steel tubes (Chiesa and Guthrie 1971, 1974).

A method for directly measuring the Lorentz number of liquid metals and alloys has been developed in the last fifteen years (Busch, Güntherodt, and Wyssmann 1972, Busch, Güntherodt, Haller, and Wyssmann 1972, 1973; Haller, Güntherodt, and Busch 1977).

8.3.2. Experimental data

At the present time, few experimental data on the thermal conductivities of liquid metals and alloys are available. In addition, the accuracy of measure-

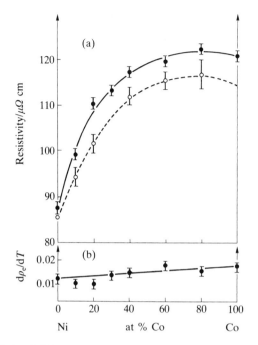

FIG. 8.7. (a) Electrical resistivity ρ_e at 1773 K, and (b) temperature coefficient $d\rho_e/dT$ for liquid nickel–cobalt alloys. ● Kita and Morita; ○ Dupree, Enderby, Newport, and Zytveld (after Kita and Morita 1984).

ments for thermal conductivities of liquid metals is relatively poor, and large discrepancies can exist among different sets of data.

Figures 8.10 and 8.11 show some experimental values of λ for liquid metals and alloys, respectively. As is obvious from these Figures, the temperature dependence of thermal conductivities is quite complex.

Experimental values of thermal conductivity of liquid metals and alloys are also listed in the paper by Viswanath and Mathur (1972).

The thermal conductivity of a metal, as well as its electrical conductivity, normally decreases on melting because of the disordered arrangement of ions in the liquid state. However, liquid metals are still much better heat conductors than non-metallic liquids. This is indicated in Fig. 8.12.

8.4. RELATIONSHIP BETWEEN ELECTRICAL CONDUCTIVITY AND THERMAL CONDUCTIVITY FOR LIQUID METALS AND ALLOYS: THE WIEDEMANN–FRANZ–LORENZ LAW

A simple theoretical relationship exists between electrical and thermal conductivities of solid metals at high temperatures:

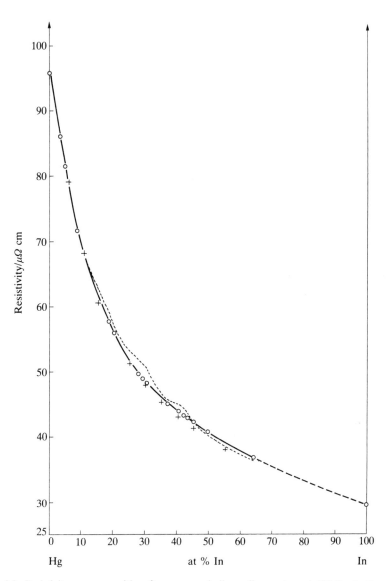

FIG. 8.8. Resistivity vs. composition for mercury–indium alloy system at 293 K. ○ Cusack, Kendall, and Fielder; + Schulz; – – – Roll and Swamy (after Cusack, Kendall, and Fielder 1964).

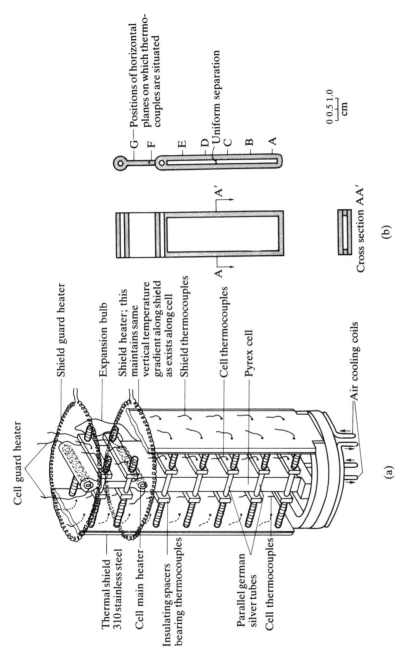

FIG. 8.9. (a) A schematic diagram of a thermal shield assembly for the longitudinal heat-flow method of thermal conductivity measurement, and (b) (Pyrex) specimen cell (after Duggin 1969).

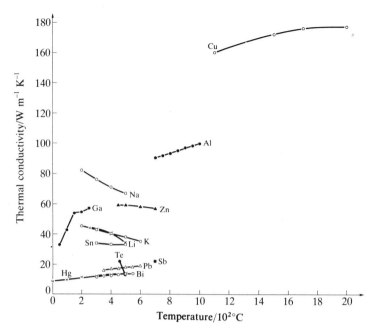

FIG. 8.10. Thermal conductivity of some pure liquid metals as a function of temperature. Data from Powell and Childs (1972).

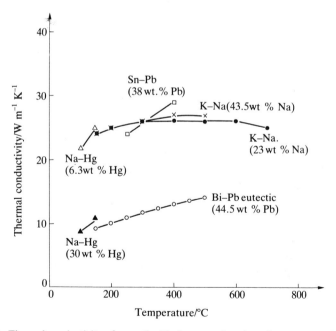

FIG. 8.11. Thermal conductivity of some liquid alloys as a function of temperature. Data from Powell and Childs (1972).

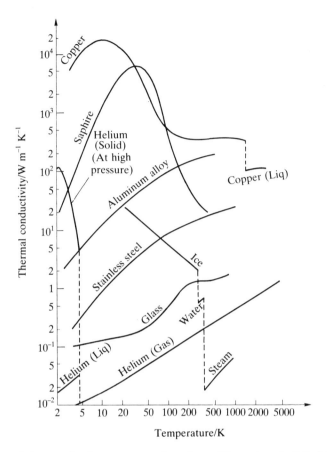

FIG. 8.12. Typical curves showing temperature dependence of thermal conductivity (after Powell and Childs 1972).

$$\frac{\lambda}{\sigma_e T} = \frac{\pi^2 k^2}{3e^2} = 2.45 \times 10^{-8} \qquad (\text{W}\,\Omega\,\text{K}^{-2}) \tag{8.5}$$

The fact that λ/σ_e ($= \lambda \rho_e$) is the same for all metals is well known as the Wiedemann–Franz–Lorenz law. The fact that this ratio is proportional to absolute temperature was discovered by Lorenz. The right-hand-side of eqn (8.5), $\pi^2 k^2/3e^2 \equiv L_0$, is called the (theoretical) Lorenz number.

Busch *et al.* (1972, 1973) and Haller *et al.* (1977) have developed a new apparatus which has allowed them to measure, with high accuracy, the Lorenz ratio or the Lorenz number of liquid metals up to 773 K. Their experimental results for the Lorenz number of liquid gallium, mercury, and tin show good agreement with the Wiedemann–Franz–Lorenz law (i.e. the free electron value), contrary to considerably larger deviations (30–100 per cent) from the law in earlier measurements[†]. Furthermore, Busch *et al.* found that liquid mercury–indium alloys also obey the Wiedemann–Franz–Lorenz law. On combining the Wiedemann–Franz–Lorenz law and electrical resistivities determined experimentally, the prediction of thermal conductivities of liquid metals and alloys is possible because reliable data on electrical conductivities of various liquid metals are available.

A few studies for predicting metals' thermal conductivities have been made (e.g. Viswanath and Mathur 1972; Ho, Powell, and Liley 1974).

[†] In earlier experiments, most determinations of the Lorenz number had been made by separate measurements of the electrical conductivity and the thermal conductivity.

APPENDIX 1

Application of some expressions to molten salts

A number of the expressions presented in this book can be applied to molten salts, provided appropriate constants of proportionality are assigned. The numerical factors are liquid type-dependent. Several examples are given in this Appendix (Morita, Iida, and Kawamoto 1985).

VELOCITY OF SOUND

Equation (4.41) can be expressed in the form

$$U_m \propto \left(\frac{T_m}{M}\right)^{\frac{1}{2}} \tag{A.1}$$

where M represents the formula weight. Figure A.1 indicates that this relationship holds reasonably well for molten salts.

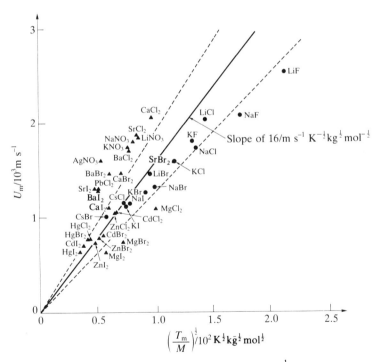

FIG. A.1. Velocity of sound in molten salts as a function of $(T_m/M)^{\frac{1}{2}}$. ● Alkali halides; broken lines denote ± 20 per cent error band. Data are from Janz (1967), Yoko, Nakano, and Ejima (1980), Ejima, Yoko, and Kato (1980), Ejima, Shimakage, Yoko, Kato, and Matsumoto (1982).

243

An alternative correlation expressed by eqn (4.44) is shown in Fig. A.2. The numerical factor 4.4×10^4 represents the best fit to the experimental data, using the method of least squares.

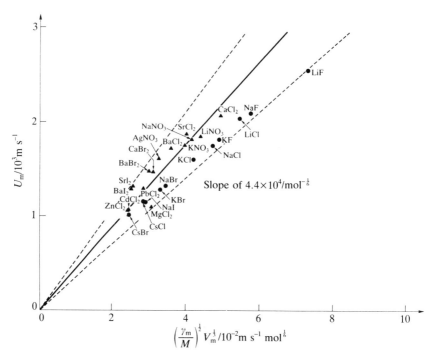

FIG. A.2. Velocity of sound in molten salts as a function of $(\gamma_m/M)^{\frac{1}{2}} V_m^{\frac{1}{3}}$. ● Alkali halides; broken lines denote ±20 per cent error band.

This relationship, i.e.

$$U_m = 4.4 \times 10^4 \left(\frac{\gamma_m}{M}\right)^{\frac{1}{2}} V_m^{\frac{1}{3}} \qquad (A.2)$$

is valid for molten salts within the ±20 per cent error band shown.

It is worth noting that a better correlation exists between U_m and $(\gamma_m/M)^{\frac{1}{2}} V_m^{\frac{1}{3}}$ than that between U_m and $(T_m/M)^{\frac{1}{2}}$, as with liquid metals.

SURFACE TENSION

The surface tension relationship given by eqn (5.35), i.e.

$$\gamma_m \propto \frac{T_m}{V_m^{\frac{2}{3}}} \qquad (A.3)$$

is shown for a large number of molten salts in Figs. A.3 to A.5. As will be evident from these figures, one can fit molten salts into three groups on the basis of this relationship:

(i)   alkali halides,
(ii) inorganic halides with the exception of the alkali halides,
(iii) carbonates, nitrates, sulphates, and miscellaneous oxides.

The need for this classification can be attributed to the types of intermolecular forces present within these inorganic compounds.

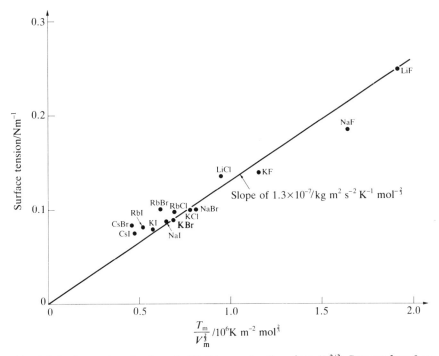

FIG. A.3. Surface tension of molten alkali halides as a function of ($T_m/V_m^{2/3}$). Data are from Janz (1967), and Ogino and Hara (1978).

VISCOSITY

Figure A.6 illustrates Andrade's viscosity formula for molten salts. As can be seen from Fig. A.6, this relationship correlates reasonably well for only the alkali halides. Other salts show large, irregular, positive deviations from the Andrade relationship of slope 1.8×10^{-7}. Such deviations from this formula (which holds for simple liquids), indicate that atomic clusters, or networks, are often formed in molten salts. As is well known, the property 'viscosity' is very sensitive to the structure of liquids. Network- or cluster-forming liquids have higher viscosity values.

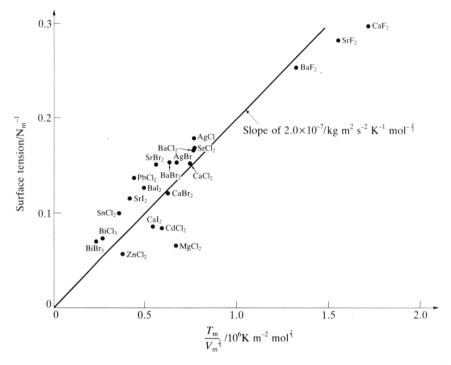

F IG. A.4. Surface tension of molten inorganic halides (excepting alkali halides) as a function of $(T_m/V_m^{2/3})$. Data are from Janz (1967), Umetsu, Kawada, Nakamura, and Ejima (1973), Ejima and Ogasawara (1977), and Ogino and Hara (1978).

THERMAL CONDUCTIVITY

In the case of non-metals, for which the electronic part of heat transport is negligible, heat is transported by vibrational motion of atoms or molecules in the liquid state. According to an early, simple theory of liquid non-metals by Osida (1939), the thermal conductivity of dielectric liquids at their melting points, λ_m, is given by

$$\lambda_m = \frac{4kv}{a} \tag{A.4}$$

where k is Boltzmann's constant, v the frequency of intermolecular vibration, and a the mean distance between two adjacent molecules. Using Lindemann's formula for the frequency, Osida has derived an explicit expression for the thermal conductivity of liquid non-metals:

$$\lambda_m = 4.1 \times 10^{-6} \frac{T_m^{\frac{1}{2}}}{M^{\frac{1}{2}} V_m^{\frac{2}{3}}} \quad \text{(in SI units)} \tag{A.5}$$

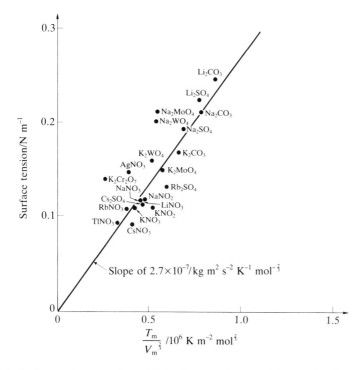

FIG. A.5. Surface tension of molten salts (carbonates, nitrates, sulphates, miscellaneous) as a function of $(T_m/V_m^{2/3})$. Data are from Janz (1967).

Subsequently, both Rao (1941) and Turnbull (1961) proposed analogous expressions to the above equation. Thus, according to Turnbull, it is more appropriate, for dissociated molten salts, to use M/n and V_m/n, rather than M and V_m, where n is the number of discrete ions. Turnbull's equation is

$$\lambda_m = 3.77 \times 10^6 \frac{T_m^{\frac{1}{2}}}{\left(\dfrac{M}{n}\right)^{\frac{1}{2}} \left(\dfrac{V_m}{n}\right)^{\frac{2}{3}}} \tag{A.6}$$

In calculating thermal conductivities using eqn (A.6), values of ionic masses, radii for discrete ions (i.e. anion and cation), as well as the number of discrete ions, are needed (Gustafsson, Halling, and Kjellander 1967). Unfortunately, there is very little information on the number of discrete ions n, so that the evaluation of n is based on assumptions that are somewhat arbitrary.

As mentioned previously, Lindemann's formula provides only rough values for v. Therefore, combining a modified Lindemann's formula by eqns (5.25) and (5.41) and eqn (A.4), we have

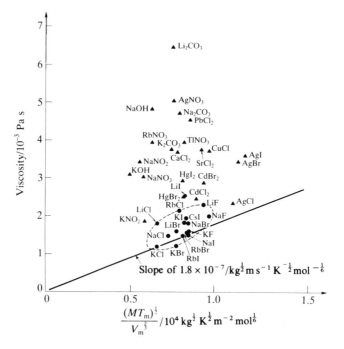

F IG. A.6. Viscosity of molten salts as a function of $(MT_m)^{\frac{1}{2}}/V_m^{\frac{2}{3}}$. ● Alkali halides. Data are from Janz (1967), Janz, Dampier, Lakshminarayanan, Lorenz, and Tomkins (1968), Janz, Gardner, Krebs, and Tomkins (1974).

$$\lambda_m = 1.2 \times 10^{-2} \left(\frac{\gamma_m}{M}\right)^{\frac{1}{2}} V_m^{-\frac{1}{3}} \qquad (A.7)$$

The numerical factor of 1.2×10^{-2} was determined so as to best fit the experimental data indicated in Fig. A.7. Osida's result is slightly improved through use of the modified Lindemann's formula. In Fig. A.8, the thermal conductivity predicted for molten CaF_2 is compared with experimental data. As can be seen, the predicted value would seem to be good. From eqns (A.2) and (A.7), we have a simple expression for the melting-point thermal conductivity:

$$\lambda_m = 2.7 \times 10^7 \frac{U_m}{V_m^{\frac{2}{3}}} \qquad (A.8)$$

Fig. A.9 indicates that a numerical factor of 3.3×10^{-7} in eqn (A.8) reproduces experimental data better than that with the factor 2.7×10^{-7}.

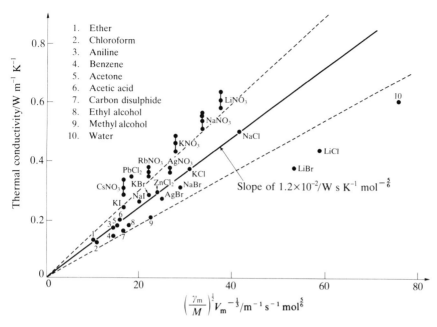

FIG. A.7. Thermal conductivity of liquid non-metals as a function of $(\gamma_m/M)^{\frac{1}{2}} V^{\frac{1}{3}}_m$. Points linked by a vertical line represent two or three different experimental values for a single salt. Broken lines denote ±25 per cent error band. Data are from Asahina and Kosaka (1980), Nagata and Goto (1985), Osida (1939), and *Rika-nenpyo* (Scientific Almanac, annually edited from Tokyo Astronomical Observatory, Maruzen Co. Ltd.). This figure was drawn with the help of Dr Tanaka of Osaka University.

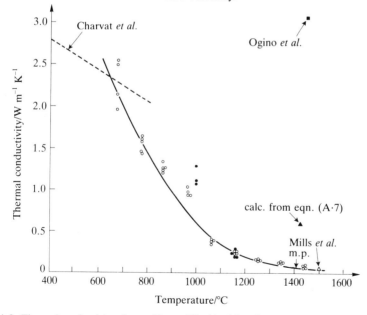

FIG. A.8. Thermal conductivity of crystalline and liquid calcium fluoride (Nagata, Susa, and Goto 1983; Nagata and Goto 1985).

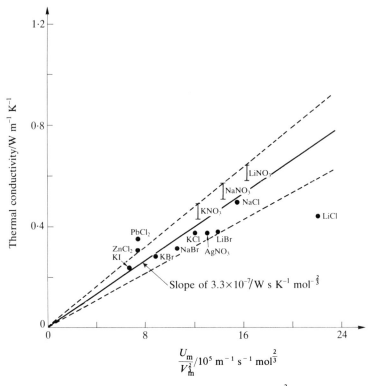

FIG. A.9. Thermal conductivity of molten salts as a function of $U_m/V_m^{\frac{2}{3}}$. The bars represent the extremes of experimental values for λ_m. Broken lines denote ± 20 per cent error band.

RELATIONSHIP BETWEEN VISCOSITY AND THERMAL CONDUCTIVITY

Combining Andrade's viscosity and Osida's thermal conductivity formulae for simple liquids, we have (Mohanty 1951)

$$\frac{\mu_m}{\lambda_m} = \frac{M}{3R} \tag{A.9}$$

where R is the gas constant. Figure A.10 indicates that this relationship is approximately true for alkali halides. Most of the molten salts excepting the alkali halides show irregular deviations from the solid line of $1/3R$. Furthermore, it should be noted that they deviate towards the upper left-hand-side of the continuous line. These deviations must stem from cluster or network formations in the molten state.

It would appear that the velocity of sound, surface tension, and thermal conductivity are not sensitive to the structure of these liquids and are mainly dominated by the number density or packing fraction.

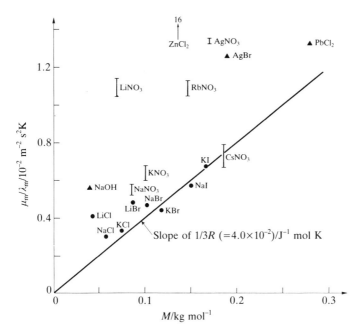

FIG. A.10. Ratio of viscosity to thermal conductivity vs. formula weight. ● Alkali halides. The bars refer to different experimental values of λ_m for a single salt. Data are from Janz (1967), Janz *et al.* (1968), Ejima, Shimakage, Yoko, Nakashima, and Takei (1979), Asahina and Kosaka (1980). Nagata and Goto (1985).

REFERENCES

Asahina, T. and Kosaka, M. (1980), *The 140th Committee, the Japan Society for the Promotion of Science (JSPS)*, Rep. No. T5, Dec. 1980 (Data are compiled by Asahina and Kosaka).

Ejima, T. and Ogasawara, M. (1977), *J. Japan Inst. Metals*, **41**, 778.

Ejima T., Shimakage, K., Yoko, T., Kato, T. and Matsumoto, S. (1982), *J. Japan Inst. Metals*, **46**, 53.

Ejima, T., Shimakage, K., Yoko, T., Nakashima, K. and Takei, K. (1979), *J. Japan Inst. Metals*, **43**, 577.

Ejima, T., Yoko, T., and Kato, T. (1980), *J. Japan Inst. Metals*, **44**, 525.

Gustafsson, S. E., Halling, N. O. and Kjellander, R. A. E. (1968), *Z. Naturforsch.*, **23a**, 682.

Janz, G. J. (1967), *Molten Salts Handbook*, Academic Press, New York.

Janz, G. J., Dampier, F. W., Lakshminarayanan, G. R., Lorenz, P. K. and Tomkins, R. P. T. (1968), *Molten Salts: Volume 1, Electrical Conductance, Density, and Viscosity Data*, Nat. Stand. Ref. Data Ser., NBS (U.S.), 15, Oct. 1968.

Janz, G. J., Gardner, G. L., Krebs, Ursula, and Tomkins, R. P. T. (1974), *Molten Salts: Volume 4, Part 1, Fluorides and Mixtures, Electrical Conductance, Density, Viscosity, and Surface Tension Data, J. Phys. Chem. Ref. Data*, **3**, No. 1.

Mohanty, S. R. (1951), *Nature*, **168**, 42.

Morita, Z., Iida, T., and Kawamoto, M. (1985), *The 140th Committee, the Japan Society for Promotion of Science (JSPS)*, Rep. No. 175, June 1985.

Nagata, K. and Goto, K. S. (1985), *The 140th Committee, the Japan Society for the Promotion of Science (JSPS)*, Rep. No. 174. June 1985.

Nagata, K., Susa, M., and Goto, K. S. (1983), *Tetsu-to-Hagané*, **69**, 1417.

Ogino, K. and Hara, S., (1978) *Tetsu-to-Hagané*, **64**, 523.

Osida, I. (1939), *Proc. Phys.-Math. Soc. Japan*, **21**, 353.

Rao, M. R. (1941), *Phys. Rev.*, **59**, 212.

Turnbull, A. G. (1961), *Aust. J. Appl. Sci.*, **12**, 324.

Umetsu, Y., Kawada, M., Nakamura, E. and Ejima, T. (1973), *J. Japan Inst. Metals*, **37**, 1139.

Yoko, T., Nakano, M. and Ejima, T. (1980), *J. Japan Inst. Metals*, **44**, 508.

APPENDIX 2

The SI (Système Internationale) Units, Physical Constants, and Conversion Factors

SI UNITS

The International System has seven fundamental units.

Physical quantity	Unit	Abbreviation
Length	metre	m
Mass	kilogram	kg
Time	second	s
Electric current	ampere	A
Thermodynamic temperature	kelvin	K
Amount of substance	mole	mol
Luminous intensity	candela	cd

Some of the most common derived units in the SI are the following.

Physical quantity	Unit	Abbreviation	Definition in fundamental units
Volume	litre	l	$m^3 \times 10^{-3}$
Force	newton	N	$m\,kg\,s^{-2}$
Energy	joule	J	$m^2\,kg\,s^{-2}\ (=N\,m)$
Power	watt	W	$m^2\,kg\,s^{-3}\ (=J\,s^{-1})$
Pressure	pascal	Pa	$m^{-1}\,kg\,s^{-2}\ (=N\,m^{-2})$
Electric charge	coulomb	C	$A\,s$
Electric potential	volt	V	$m^2\,kg\,s^{-3}\,A^{-1}\ (=W\,A^{-1})$
Electric resistance	ohm	Ω	$m^2\,kg\,s^{-3}\,A^{-2}\ (=V\,A^{-1})$
Frequency	hertz	Hz	s^{-1}

Multiples or decimal fractions of the basic SI units are designated by prefixes.

Fraction	Prefix	Symbol	Multiple	Prefix	Symbol
10^{-1}	deci	d	10	deka	da
10^{-2}	centi	c	10^2	hecto	h
10^{-3}	milli	m	10^3	kilo	k
10^{-6}	micro	μ	10^6	mega	M
10^{-9}	nano	n	10^9	giga	G
10^{-12}	pico	p	10^{12}	tera	T
10^{-15}	femto	f	10^{15}	peta	P
10^{-18}	atto	a	10^{18}	exa	E

PHYSICAL CONSTANTS

Gravitational acceleration
at sea level and equator $\qquad g = 9.806\ m\,s^{-2}$

Avogadro's number $N_A = 6.0221367 \times 10^{23}$ (atoms) mol^{-1}
Boltzmann's constant $k = 1.380658 \times 10^{-23}$ JK^{-1}
Electron charge $e = 1.60217733 \times 10^{-19}$ C
Faraday's constant $F = 96485.3$ C mol^{-1}
Gas constant $R = 8.31451$ JK^{-1} mol^{-1}
                                       $= 0.082058$ litre atm K^{-1} mol^{-1}
                                       $= 1.9872$ cal K^{-1} mol^{-1}
Mass of electron $m_e = 9.1093897 \times 10^{-31}$ kg
Planck's constant $h = 6.6260755 \times 10^{-34}$ J s
Speed of light $c = 2.9979246 \times 10^{8}$ m s^{-1}

CONVERSION FACTORS

1 electron volt (eV) $= 1.602189 \times 10^{-19}$ J
1 calorie $= 4.184$ J
1 litre atm $= 101.325$ J
1 atm $= 1.01325 \times 10^{5}$ Pa (or N m^{-2}) $= 1.01325$ bar $= 760$ torr (or mmHg)
$2.303\,RT = 19.148$ k J mol^{-1} at 1000 K
1 k J mol$^{-1} = 83.593$ cm^{-1}
$0\,K = -273.15°C$

A 'Handbook' on the physical properties of metallurgical materials at high temperatures will be published in 1988 from the 140th Committee the Japan Society for the Promotion of Science (JSPS). Experimental data for the various physical properties of liquid metals, alloys, and slags, over about the last two decades, will be compiled. (The Handbook will contain only experimental data).

REFERENCES

Abowitz, G. (1977). *Scripta Met.* **11**, 353.

Adachi, A., Ogino, K., and Kawasaki, S. (1963). *Technol. Repts. Osaka Univ.* **13**, 411.

Adachi, A., Morita, Z., Kitaura, M., and Demukai, N. (1970). *Technol. Repts. Osaka Univ.* **20**, 67.

Adams, P. D. and Leach, J. S. L1. (1967). *Phys. Rev.* **156**, 178.

Alder, B. J. and Wainwright, T. E. (1957). *J. Chem. Phys.* **27**, 1208.

Alder, B. J. and Wainwright, T. E. (1959). *J. Chem. Phys.* **31**, 459.

Alder, B. J. and Wainwright, T. E. (1960). *J. Chem. Phys.* **33**, 1439.

Alder, B. J. and Wainwright, T. E. (1962). *Phys. Rev.* **127**, 359.

Alfred, L. C. R. and March, N. H. (1955). *Phil. Mag.* **46**, 759.

Alfred, L. C. R. and March, N. H. (1956). *Phys. Rev.* **103**, 877.

Allen, B. C. (1963). *Trans. Met. Soc. AIME* **227**, 1175.

Allen, B. C. (1972a). *Liquid metals: chemistry and physics* (ed. S. Z. Beer). Marcel Dekker, New York, p. 182 (Table 3).

—— (1972b) —— p. 184 (Table 4).

—— (1972c) —— p. 186 (Table 5).

—— (1972d) —— p. 171 (Fig. 4).

—— (1972e) —— p. 197 (Fig. 11).

—— (1972f) —— Ch. 4.

Allnatt, A. R. and Rice, S. A. (1961). *J. Chem. Phys.* **34**, 2156.

Andrade, E. N. da C. (1934). *Phil. Mag.* **17**, 497, 698.

Andrade, E. N. da C. and Chiong, Y. S. (1936). *Proc. Phys. Soc.* **48**, 247.

Andrade, E. N. da C. and Dobbs, E. R. (1952). *Proc. R. Soc.* **A211**, 12.

Arsentiev, P. P., Vinogradov, B. G., and Lisichkii, B. S. (1974) 7. *Izv. Vuzov. Chern. Met.* 181.

Ascarelli, P. (1968). *Phys. Rev.* **173**, 271.

Ashcroft, N. W. and Lekner, J. (1966). *Phys. Rev.* **145**, 83.

Atterton, D. V. and Hoar, T. P. (1951). *Nature* **167**, 602.

Baes, C. F. and Kellogg, H. H. (1953). *Trans. AIME* **197**, 643.

Bansal, R. (1973). *J. Phys. C* **6**, 1204, 3071.

Bardeen, J. and Pines, D. (1955). *Phys. Rev.* **99**, 1140.

Barker, J. A. and Henderson, D. (1981). *Sci. Am.* **245**, 130.

Barrett, C. S. (1966). Structure of metals, Table A-6, McGraw-Hill, New York.

Bashforth, F. and Adams, J.C. (1883). *An attempt to test the theories of capillary action.* Cambridge University Press.

Basin, A. S. and Solov'ev, A. N. (1967). *Zhur. Priklad. Mekh. Tekhn. Fiziki* No. 6, 83.

Bell, C. F. and Lott, K. A. K. (1963). *Modern approach to inorganic chemistry.* p. 195, Butterworths, London.

Belton, G. R. (1976). *Met. Trans.* **7B**, 35.

Belton, J. W. and Evans, M. G. (1945). *Trans. Faraday Soc.* **41**, 1.

Bernal, J. D. (1959). *Nature* **183**, 141.

Bernard, G. and Lupis, C. H. P. (1971). *Met. Trans.* **2**, 555.

Berthou, P. E. and Tougas, R. (1968). *J. Less-Common Metals* **16**, 465.

Beyer, R. T. and Ring, E. M. (1972). *Liquid metals: chemistry and physics* (ed. S. Z. Beer). Ch. 9, Marcel Dekker, New York.

Bhatia, A. B. and March, N. H. (1978a). *J. Chem. Phys.* **68**, 1999.

Bhatia, A. B. and March, N. H. (1978*b*). *J. Chem. Phys.* **68**, 4651.

Bird, R. B., Stewart, W. E., and Lightfoot, E. N. (1960). *Transport phenomena* Ch. 1. John Wiley & Sons, New York.

Bockris, J. O'M., White, J. L., and Mackenzie, J. D. (1959). *Physicochemical measurements at high temperatures.* Butterworths, London.

Bohm, D. and Staver, T. (1951). *Phys. Rev.* **84**, 836.

Born, M. and Green, H. S. (1947). *Proc. R. Soc.* **A190**, 455.

Brager, A. and Schuchowitzky, A. (1946). *Acta Physicochimica U.R.S.S.* **21**, 13.

Braunbek, W. (1932). *Z. Phys.* **73**, 312.

Breitling, S. M. and Eyring, H. (1972). *Liquid metals: chemistry and physics* (ed. S. Z. Beer). Ch. 5, Marcel Dekker, New York.

Broome, E. F. and Walls, H. A. (1968). *Trans. Met. Soc. AIME* **242**, 2177.

Broome, E. F. and Walls, H. A. (1969). *Trans. Met. Soc. AIME* **245**, 739.

Brown, D. S. and Tuck, D. G. (1964). *Trans. Faraday Soc.* **60**, 1230.

Buff, F. P. (1952). *Z. Elektrochem.* **56**, 311.

Busch, G., Güntherodt, H.-J., and Wyssmann, P. (1972). *Phys. Lett.* **39A**, 89.

Busch, G., Güntherodt, H.-J., Haller, W., and Wyssmann, P. (1972). *Phys. Lett.* **41A**, 29.

Busch, G., Güntherodt, H.-J., Haller, W., and Wyssmann, P. (1973). *Phys. Lett.* **43A**, 225.

Butler, J. A. V. (1932). *Proc. R. Soc.* **A135**, 348.

Calderon, F. P., Sano, N., and Matsushita, Y. (1971). *Met. Trans.* **2**, 3325.

Calverley, A. (1957). *Proc. Phys. Soc.* **70B**, 1040.

Careri, G. and Paoletti, A. (1955). *Nuovo Cimento* **2**, 574.

Careri, G., Paoletti, A., and Vicentini, M. (1958). *Nuovo Cimento* **10**, 1088.

Cavalier, G. (1963). *Compt. Rend.* **256**, 1308.

Chapman, T. W. (1966). *Mater. Sci. Eng.* **1**, 65.

Chiesa, F. and Guthrie, R. I. L. (1971). *Met. Trans.* **2**, 2833.

Chiesa, F. and Guthrie, R. I. L. (1974). *Trans. ASME, J. Heat Transfer*, **96**, *C 3*, 377.

Chiong, Y. S. (1936). *Proc. R. Soc.* **A157**, 264.

Cocking, S. J. (1969). *J. Phys. C* **2**, 2047.

Cohen, M. H. and Turnbull, D. (1959). *J. Chem. Phys.* **31**, 1164.

Cook, A. H. (1961). *Phil. Trans. R. Soc.* **A254**, 125.

Crawley, A. F. (1968). *Trans. Met. Soc. AIME* **242**, 2309.

Crawley, A. F. (1974). *Int. Met. Rev.* **19**, 32.

Crawley, A. F. and White, D. W. G. (1968). *Trans. Met. Soc. AIME* **242**, 1483.

Croxton, C. A. and Ferrier, R. P. (1971*a*). *Phil. Mag.* **24**, 489, 493.

Croxton, C. A. and Ferrier, R. P. (1971*b*). *J. Phys. C* **4**, 1909, 1921, 2433, 2447.

Cusack, N. and Enderby, J. E. (1960). *Proc. Phys. Soc.* **75**, 395.

Cusack, N., Kendall, P., and Fielder, M. (1964). *Phil. Mag.* **10**, 871.

Darken, L. S. and Gurry, R. W. (1953). *Physical chemistry of metals.* Ch. 18, McGraw-Hill, New York.

Davies, V. de L. and West, J. M. (1963–4). *J. Inst. Metals* **92**, 208.

Davis, K. G. (1966). *Can. Met. Q.* **5**, 245.

Davis, H. T. and Palyvos, J. A. (1967). *J. Chem. Phys.* **46**, 4043.

Depuydt, P. J. and Parlee, N. A. D. (1972). *Met. Trans.* **3**, 525.

Dillon, I. G., Nelson, P. A., and Swanson, B. S. (1966). *J. Chem. Phys.* **44**, 4229.

Djemili, B., Martin-Garin, L., Martin-Garin, R., and Desré, P. (1981). *J. Less-common Metals* **79**, 29.

Djemili, B., Martin-Garin, L., Martin-Garin, R., and Hicter, P. (1980). Proc. 4th Int. Conf. on Liquid and Amorphous Metals, Grenoble, 7–11 July, in *J. Phys. (Paris), Colloq.* **41**, C8–363.

Döge, G. (1966). *Z. Naturforsch.* **21a**, 266.

Dorsey, N. E. (1928). *J. Washington Acad. Sci.* **18**, 505.

Duggin, M. J. (1969). *Thermal conductivity* (ed. C. Y. Ho and R. E. Taylor). Proc. 8th Conf., TRRC, Purdue University, p. 727, Plenum Press.

Edwards, J. B., Hucke, E. E., and Martin, J. J. (1968). *Met. Rev.* **120** (Part II), 13.

Egelstaff, P. A. (1967a). *An introduction to the liquid state.* Ch. 5, Academic Press, London.

—— (1967b) —— p. 17.

—— (1967c) —— Ch. 12.

—— (1967d) —— Ch. 14.

—— (1967e) —— Ch. 10.

Egelstaff, P. A. and Widom, B. (1970). *J. Chem. Phys.* **53**, 2667.

Einstein, A. (1911). *Ann. d. Physik* **34**, 170.

Eisenstein, A. and Gingrich, N. S. (1942). *Phys. Rev.* **62**, 261.

Ejima, T., Inagaki, N., and Kameda, M. (1966). *Trans. JIM* **7**, 133.

El-Mehairy, A. E. and Ward, R. G. (1963). *Trans. Met. Soc. AIME* **227**, 1226.

Ershov, G. S., Kasatkin, A. A., and Gavrilin, I. V. (1976) 2. *Izv. Akad. Nauk SSSR. Metally* p. 98.

Evans, R., Greenwood, D. A., and Lloyd, P. (1971). *Phys. Lett.* **35A**, 57.

Evans, R., Gyorffy, B. L., Szabo, N., and Ziman, J. M. (1973). *Proc. 2nd Int. conf. on liquid metals, Tokyo* (ed. S. Takeuchi). p. 319 Taylor and Francis, London.

Ewing, C. T., Grand, J. A., and Miller, R. R. (1951). *J. Am. Chem. Soc.* **73**, 1168.

Eyring, H., Henderson, D., Stover, B. J., and Eyring, E. M. (1964). *Statistical mechanics and dynamics.* John Wiley & Sons, New York.

Faber, T. E. (1972a). *Introduction to the theory of liquid metals.* p. 165, Cambridge University Press.

—— (1972b) —— p. 174.

—— (1972c) —— Ch. 5.

Faber, T. E. and Ziman, J. M. (1965). *Phil. Mag.* **11**, 153.

Filippov, S. I., Kazakov, N. B., and Pronin, L. A. (1966). *Izv. VUZ Chern. Met.* **9**(3), 8.

Fisher, H. J. and Phillips, A. (1954). *Trans. AIME* **200**, 1060.

Fisk, S. and Widom, B. (1969). *J. Chem. Phys.* **50**, 3219.

Fordham, S. (1948). *Proc. R. Soc.* **A194**, 1.

Forster, D., Martin, P. C., and Yip, S. (1968). *Phys. Rev.* **170**, 155, 160.

Fort, R. J. and Moore, W. R. (1966). *Trans. Faraday Soc.* **62**, 1112.

Fowler, R. H. (1937). *Proc. R. Soc.* **A159**, 229.

Fowler, R. H. and Guggenheim, E. A. (1965). *Statistical thermodynamics.* Ch. V, Cambridge University Press.

Fraser, M. E., Lu, W.-K., Hamielec, A. E., and Murarka, R. (1971). *Met. Trans.* **2**, 817.

Frenkel, J. (1917). *Phil. Mag.* **33**, 297.

Frenkel, J. (1946). *Kinetic theory of liquids.* Oxford University Press, London.

Friedel, J. (1952). *Phil. Mag.* **43**, 153.

Frohberg, M. G. and Cakici, T. (1977). *Arch. Eisenhüttenw.* **48**, 145.

Frohberg, M. G. and Weber, R. (1964). *Arch. Eisenhüttenw.* **35**, 885.

Furukawa, K. (1960). *Sci. Rep. Res. Inst. Tohoku Univ.* **12A**, 368.

Ganovici, I. and Ganovici, L. (1970). *Rev. Roum. Chim.* **15**, 213.

Gebhardt, E., Becker, M., and Dorner, S. (1953). *Z. Metallk.* **44**, 573.

Gebhardt, E., Becker, M., and Dorner, S. (1954). *Z. Metallk.* **45**, 83.

Gebhardt, E. and Detering, K. (1959). *Z. Metallk.* **50**, 379.

Gebhardt, E. and Wörwag, G. (1951). *Z. Metallk.* **42**, 358.

Giedt, W. H. (1971). *Thermophysics.* P. 58, Van Nostrand Reinhold Company, New York.

Gitis, M. B. and Mikhailov, I. G. (1966). *Sov. Phys. Acoust.* **12**, 14.

Gitis, M. B. and Mikhailov, I. G. (1967). *Sov. Phys. Acoust.* **13**, 251.

Gitis, M. B. and Mikhailov, I. G. (1968). *Sov. Phys. Acoust.* **13**, 473.

Glasstone, S., Laidler, K. J., and Eyring, H. (1941). *The theory of rate processes*, Ch. IX, McGraw-Hill, New York.

Gogate, D. V. and Kothari, D. S. (1935). *Phil. Mag.* **20**, 1136.

Goto, R., Hirai, N., and Hanai, T. (1964). *Rheology and its applications.* (in Japanese). p. 47, Kyoritsu Press, Tokyo.

Gourtsoyannis, L. *Kinetics of compound gas absorption by liquid iron and nickel.* Ph.D. Thesis, McGill Univ., 1978.

Grace, R. E. and Derge, G. (1955). *Trans. AIME* **203**, 839.

Grosse, A. V. (1961a). *J. Inorg. Nucl. Chem.* **22**, 23.

Grosse, A. V. (1961b). *J. Inorg. Nucl. Chem.* **23**, 333.

Grosse, A. V. (1964). *J. Inorg. Nucl. Chem.* **26**, 1349.

Guggenheim, E. A. (1945). *Trans. Faraday Soc.* **41**, 150.

Halden, F. A. and Kingery, W. D. (1955). *J. Phys. Chem.* **59**, 557.

Haller, W., Güntherodt, H.-J., and Busch, G. (1977). *Liquid metals 1976.* P. 207, Inst. Phys. Conf. Ser. No. 30, Bristol.

Handbook of chemistry and physics (1967–8) (48th edn ed R. C. Weast). F 143, Chemical Rubber Co., Cleveland.

Hansen, M. (1958). *Constitution of binary alloys.* p. 337, p. 624 McGraw-Hill, New York.

Harashima, A. (1953). *J. Phys. Soc. Japan* **8**, 343.

Harkins, W. D. and Brown, F. E. (1919). *J. Am. Chem. Soc.* **41**, 499.

Harrison, W. A. (1966). *Pseudopotentials in the theory of metals.* Benjamin, New York.

Helfand, E. (1960). *Phys. Rev.* **119**, 1.

Helfand, E. (1961). *Phys. Fluids* **4**, 681.

Helfand, E. and Rice, S. A. (1960). *J. Chem. Phys.* **32**, 1642.

Henderson, J. and Yang, L. (1961). *Trans. Met. Soc. AIME* **221**, 72.

Hicter, P., Durand, F., and Bonnier, E. (1971). *J. Chim. Phys. (France)* **68**, 804.

Hiemstra, S., Prins, D., Gabrielse, G., and Zytveld, J. B. Van (1977). *Phys. Chem. Liquids* **6**, 271.

Ho, C. Y., Powell, R. W., and Liley, P. E. (1974). *Thermal conductivity of elements: a comprehensive review.* Vol. 3, Sup. No. 1, pp. 1–11, the American Chemical Society and the American Institute of Physics for the National Bureau of Standards.

Hoar, T. P. and Melford, D. A. (1957). *Trans. Faraday Soc.* **53**, 315.

Hoffman, R. E. (1952). *J. Chem. Phys.* **20**, 1567.

Hogness, T. R. (1921). *J. Am. Chem. Soc.* **43**, 1621.

Hoover, W. G. and Ree, F. H. (1967). *J. Chem. Phys.* **47**, 4873.

Hoover, W. G. and Ree, F. H. (1968). *J. Chem. Phys.* **49**, 3609.

Hopkins, M. R. and Toye, T. C. (1950). *Proc. Phys. Soc.* **B63**, 773.

Hsieh, M. and Swalin, R. A. (1974). *Acta Met.* **22**, 219.

Huang, K. and Wyllie, G. (1949). *Proc. Phys. Soc.* **A62**, 180.

Hudson, S. and Andersen, H. C. (1978). *J. Chem. Phys.* **69**, 2323.

Iida, T. (1970). Dr. Eng. Thesis, Tohthoku University.

Iida, T. and Morita, Z. (1978). The 140th Committee, the Japan Society for Promotion of Science (JSPS), Rep. No. 69, Dec. 1978.

Iida, T. and Morita, Z. (1980). *Bull. Japan Inst. Metals* **19**, 655.

Iida, T., Fukase, S., and Morita, Z. (1981). *Bull. Japan Inst. Metals* **20**, 264.

Iida, T., Guthrie, R. I. L., and Morita, Z. (1982). *Proc. phys. chem. iron steel making, Toronto*, III-25.

Iida, T., Kijima, H., and Morita, Z. (1983). Tetsu-to-Hagané **69**, S942.

Iida, T., Kumada, T., Washio, M., and Morita, Z. (1980). *J. Japan Inst. Metals* **44**, 1392.

Iida, T., Morita, Z., and Chikazawa, B. (1978). Tetsu-to-Hagané **64**, S629.

Iida, T., Morita, Z., and Takeuchi, S. (1975). *J. Japan Inst. Metals* **39**, 1169.

Iida, T., Ueda, M., and Morita, Z. (1976). Tetsu-to-Hagané **62**, 1169.

Iida, T., Kasama, A., Misawa, M., and Morita, Z. (1974). *J. Japan Inst. Metals* **38**, 177.

Iida, T., Kasama, A., Morita, Z., Okamoto, I., and Tokumoto, S. (1973). *J. Japan Inst. Metals* **37**, 841.

Iida, T., Satoh, A., Ishiura, S., Ishiguro, S., and Morita, Z. (1980). *J. Japan Inst. Metals* **44**, 443.

Iida, T., Washio, M., Kumada, T., and Morita, Z. (1980). *High Temp. Soc. Japan* **6**, 197.

Inouye, M. and Choh, T. (1968). *Trans. ISIJ* **8**, 134.

Irons, G. A. and Guthrie, R. I. L. (1981). *Can. Met. Q.* **19**, 381.

Isherwood, S. P. and Orton, B. R. (1972). *J. Phys. C* **5**, 2985.

Johnson, M. D., Hutchinson, P., and March, N. H. (1964). *Proc. R. Soc.* **A282**, 283.

Jones, W. R. D. and Bartlett, W. L. (1952–3). *J. Inst. Metals* **81**, 145.

Kasama, A., Iida, T., and Morita, Z. (1976). *J. Japan Inst. Metals* **40**, 1030.

Kawai, Y. and Mori, K. (1979). *Studies on metallic melts in metallurgical reactions* (ed. M. Shimoji). P. 50, March 1979.

Kawakami, M. and Goto, K. S. (1976). *Trans. ISIJ* **16**, 204.

Keskar, A. R. and Hruska, S. J. (1970). *Met. Trans.* **1**, 2357.

Kijima, H. (1983). B. Eng. Thesis, Osaka Univ.

Kingery, W. D. (1959). *Property measurements at high temperatures.* P. 370, John Wiley & Sons.

Kingery, W. D. and Humenik, M. (1953). *J. Phys. Chem.* **57**, 359.

Kirkwood, J. G. and Buff, F. P. (1949). *J. Chem. Phys.* **17**, 338.

Kirshenbaum, A. D. and Cahill, J. A. (1960). *J. Inorg. Nucl. Chem.* **14**, 283.

Kirshenbaum, A. D. and Cahill, J. A. (1962a). *Trans. ASM* **55**, 844.

Kirshenbaum, A. D. and Cahill, J. A. (1962b). *Trans. ASM* **55**, 849.

Kirshenbaum, A. D., Cahill, J. A., and Grosse, A. V. (1961). *J. Inorg. Nucl. Chem.* **22**, 33.

Kirshenbaum, A. D., Cahill, J. A., and Grosse, A. V. (1962). *J. Inorg. Nucl. Chem.* **24**, 333.

Kita, Y. and Morita, Z. (1984). *Liquid and amorphous metals*, Part II, (eds. C. N. J. Wagner and W. L. Johnson). P. 1079, North-Holland Physics Publishing, Amsterdam.

Kita, Y., Oguchi, S., and Morita, Z. (1978). Tetsu-to-Hagané **64**, 711.

Kita, Y., Zeze, M., and Morita, Z. (1982). *Trans. ISIJ* **22**, 571.

Kitajima, M., Saito, K., and Shimoji, M. (1976). *Trans. JIM* **17**, 582.

Kittel, C. (1971). *Introduction to solid state physics* (4th edn). P. 96, John Wiley & Sons.

Kleinschmit, P. and Grothe, K. H. (1970). *Z. Metallk.* **61**, 378.

Kleppa, O. J. (1950). *J. Chem. Phys.* **18**, 1331.

Kleppa, O. J. (1960). *J. Phys. Chem.* **64**, 1542.

Kleppa, O. J., Kaplan, M., and Thalmayer, C. E. (1961). *J. Phys. Chem.* **65**, 843.

Knappwost, A. (1948). *Z. Metallk.* **39**, 314.

Knappwost, A. (1952). *Z. Phys. Chem.* **200**, 81.

Knight, F. W. (1961). *Plutonium* 1960. (ed. E. Grison, W. B. H. Lord, and R. D. Fowler) P. 684, Cleaver-Hume Press, London.

Krieger, W. and Trenkler, H. (1971). *Arch. Eisenhüttenw.* **42**, 175, 685.

Kubaschewski, O. and Alcock, C. B. (1979a). *Metallurgical thermochemistry* (5th Edn, revised and enlarged). Pergamon Press, Oxford, p. 268 (Table A).

—— (1979b) —— p. 326 (Table B); p. 358 (Table D).

—— (1979c) —— p. 55.

—— (1979d) —— p. 358 (Table D).

—— (1979e) —— p. 336 (Table C1).

—— (1979f) ——, Ch. II.

Kubo, R. (1957). *J. Phys. Soc. Japan* **12**, 570, 1203.

Kubo, R., Yokota, M., and Nakajima, S. (1957). *J. Phys. Soc. Japan*, **12**, 1203.

Kunze, H. D. (1973). *Arch. Eisenhüttenw.* **44**, 71, 173.

Landau, L. D. and Lifshitz, E. M. (1958). *Statistical physics*. Pergamon Press, Oxford.

Lang, G. (1973). *J. Inst. Metals* **101**, 300.

Lange, W., Pippel, W., and Bendel, F. (1959). *Z. Phys. Chem.* **212**, 238.

Larsson, K. E. (1968). *Neutron inelastic scattering*. International Atomic Energy Agency (IAEA), Vienna, Vol. I, 397

Larsson, K. E., Dahlborg, U., and Jovic, D. (1965). *Inelastic Scattering of Neutrons* IAEA, Vienna, Vol. II, 117.

Larsson, S. J., Roxbergh, C., and Lodding, A. (1972). *Phys. Chem. Liquids* **3**, 137.

Leak, V. G. and Swalin, R. A. (1964). *Trans. Met. Soc. AIME* **230**, 426.

Lihl, F., Nachtigall, E., and Schwaiger, A. (1968). *Z. Metallk.* **59**, 213.

Lindemann, F. A. (1910). *Phys. Z.* **11**, 609.

Lodding, A. (1956). *Z. Naturforsch.* **11a**, 200.

Longuet-Higgins, H. C. and Pople, J. A. (1956). *J. Chem. Phys.* **25**, 884.

Lucas, L. D. (1970). *Techniques of metals research* (ed. R. F. Bunshah) Vol. 4, Part 2, p. 219, Interscience Publishers, New York.

Ma, C. H. and Swalin, R. A. (1960). *Acta Met.* **8**, 388.

Macedo, P. B. and Litovitz, T. A. (1965). *J. Chem. Phys.* **42**, 245.

Mackenzie, J. D. (1956) *Rev. Sci. Instr.* **27**, 297.

Mackenzie, J. D. (1959). *J. Phys. Chem.*, **63**, 1875.

Maier, U. and Steeb, S. (1973). *Phys. cond. Matter* **17**, 1.

March, N. H. and Tosi, M. P. (1976). *Atomic dynamics in liquids*. Ch. 10, MacMillan, London.

Marcus, Y. (1977). *Introduction to liquid state chemistry*. Ch. 8, John Wiley & Sons, London.

Martinez, J. and Walls, H. A. (1973). *Met. Trans.* **4**, 1419.

Matsuda, H. and Hiwatari, Y. (1973). *Cooperative phenomena* (eds. H. Haken and M. Wagner). Springer-Verlag.

Mayer, S. W. (1963). *J. Phys. Chem.*, **67**, 2160.

McGonigal, P. J. (1962). *J. Phys. Chem.* **66**, 1686.
McGonigal, P. J. and Grosse, A. V. (1963). *J. Phys. Chem.* **67**, 924.
Melford, D. A. and Hoar, T. P. (1956–7). *J. Inst. Metals* **85**, 197.
Menz, W. and Sauerwald, F. (1966). *Acta Met.* **14**, 1617.
Mera, Y., Kita, Y., and Adachi, A. (1972). *Technol. Repts. Osaka Univ.* **22**, 445.
Meyer, R. E. (1961) *J. Phys. Chem.* **65**, 567.
Meyer, R. E. and Nachtrieb, N. H. (1955). *J. Chem. Phys.* **23**, 1851.
Mittag, U. and Lange, K. W. (1975). *Arch. Eisenhüttenw.* **46**, 249.
Moelwyn-Hughes, E. A. (1961a). *Physical chemistry.* Ch. XIX. Pergamon Press, Oxford.
—— (1961b). —— Ch. XVII.
—— (1961c) —— Chs I, II.
Monma, K. and Suto, H. (1961). *J. Japan Inst. Metals* **25**, 65.
Morgan, D. W. and Kitchener, J. A. (1954). *Trans. Faraday Soc.* **50**, 51.
Mori, K., Kishimoto, M., Shimose, T., and Kawai, Y. (1975). *J. Japan Inst. Metals* **39**, 1301.
Morita, Z. and Iida, T. (1982). *Proc. 1st China-Japan symposium on science and technology of iron and steel.* (This paper was based on Yanagitani's work: Yanagitani, A. (1981). M. Eng. Thesis, Osaka University).
Morita, Z., Iida, T., and Kasama, A. (1976). *Bull. Japan Inst. Metals* **15**, 743.
Morita, Z., Iida, T., Kawamoto, M., and Mōri, A. (1984). Tetsu-to-Hagané **70**, 1242.
Morita, Z. Iida, T., and Matsumoto, Y. (1985). *The 140th Committee the Japan Society for the Promotion of Science (JSPS),* Rep. No. 177, Dec. 1985.
Morita, Z., Iida, T., and Ueda, M. (1976). The 140th Committee the Japan Society for the Promotion of Science (JSPS), Rep. No. 45, Dec. 1976.
Morita, Z., Iida, T., and Ueda, M. (1977). *Liquid metals 1976.* P. 600, Inst. Phys. Conf. Ser. No. 30, Bristol.
Morita, Z., Ogino, Y., Iba, T., Maehana, T., and Adachi, A. (1970). Tetsu-to-Hagané **56**, 1613.
Morita, Z., Ogino, Y., Kaitoh, H., and Adachi, A. (1970). *J. Japan Inst. Metals* **34**, 248.
Mott, N. F. (1934). *Proc. R. Soc.* **A146**, 465.
Mott, N. F. (1936), *Proc. Cambridge Phil Soc.* **32**, 281.
Murarka, R. N., Lu, W-K., and Hamielec, A. E. (1971). *Met. Trans.* **2**, 2949.
Murarka, R. N., Lu, W-K., and Hamielec, A. E. (1975). *Can. Met. Q.* **14**, 111.
Murday, J. S. and Cotts, R. M. (1968). *J. Chem. Phys.* **48**, 4938.
Murday, J. S. and Cotts, R. M. (1971). *Z. Naturforsch.* **26a**, 85.
Nachtrieb, N. H. (1967). *Adv. Phys.* **16**, 309.
Nachtrieb, N. H. (1972). *Liquid metals: chemistry and physics* (ed. S. Z. Beer) p. 509, Marcel Dekker, New York.
Nachtrieb, N. H. (1976), *Ber. Bunsenges Phys. Chem.* **80**, 678.
Nachtrieb, N. H., Fraga, E., and Wahl, C. (1963). *J. Phys. Chem.* **67**, 2353.
Nachtrieb, N. H. and Petit, J. (1956). *J. Chem. Phys.* **24**, 746.
Norden, A. and Lodding, A. (1967). *Z. Naturforsch.* **22a**, 215.
Ocken, H. and Wagner, C. N. J. (1966). *Phys. Rev.* **149**, 122.
Ofte, D. and Wittenberg, L. J. (1963). *Trans. Met. Soc. AIME* **227**, 706.
Ogino, Y., Borgmann, F. O., and Frohberg, M. G. (1973). *J. Japan Inst. Metals* **37**, 1230.
Ogino, K., Nishiwaki, A., and Hosotani, Y. (1984a). *J. Japan Inst. Metals* **48**, 996.
Ogino, K., Nishiwaki, A., and Hosotani, Y. (1984b). *J. Japan Inst. Metals* **48**, 1004.

Ogino, K., Nogi, K., and Yamase, O. (1980). Tetsu-to-Hagané **66**, 179.

Okajima, Y. and Shimoji, M. (1972). *Trans. JIM.* **13**, 255.

Olsen, D. A. and Johnson, D. C. (1963). *J. Phys. Chem.* **67**, 2529.

Ono, Y. (1977). Tetsu-to-Hagané **63**, 1350.

Ono, Y., Hirayama, K. and Furukawa, K. (1974). Tetsu-to-Hagané **60**, 2110.

Ono, Y. and Yagi, T. (1972). *Trans. ISIJ.* **12**, 314.

Onoprienko, G. I., Kuzmenko, P. P. and Kharkov, E. I. (1966). *Fiz. Meta. Metalloved.* **22**, 791.

Oriani, R. A. (1950). *J. Chem. Phys.* **18**, 575.

Osida, I. (1939). *Proc. Phys.-Math. Soc. Japan* **21**, 353.

Otsuka, S., Katayama, I., and Kozuka, Z. (1971). *Trans. JIM.* **12**, 442.

Otsuka, S. and Kozuka, Z. (1977). *Trans. JIM.* **18**, 690.

Ott, A. and Lodding, A. (1965). *Z. Naturforsch.* **20a**, 1578.

Ozelton, M. W. and Swalin, R. A. (1968). *Phil. Mag.* **18**, 441.

Parish, R. V. (1977). *The metallic elements.* P. 208, Longman.

Pasternak, A. D. (1972). *Phys. Chem. Liquids.* **3**, 41.

Perkins, R. H., Geoffrion, L. A., and Biery, J. C. (1965). *Trans. Met. Soc. AIME* **233**, 1703.

Petit, J. and Nachtrieb, N. H. (1956). *J. Chem. Phys.* **24**, 1027.

Powell, R. L. and Childs, G. E. (1972). *American Institute of Physics handbook.* pp. 4–142, McGraw-Hill.

Predel, B. and Eman, A. (1969). *Mater. Sci. Eng.* **4**, 287.

Protopapas. P., Andersen, H. C., and Parlee, N. A. D. (1973). *J. Chem. Phys.* **59**, 15.

Protopapas, P. and Parlee, N. A. D. (1976). *High Temp. Sci*, **8**, 141.

Reiss, H., Frisch, H. L., and Lebowitz, J. L. (1959). *J. Chem. Phys.* **31**, 369.

Reynik, R. J. (1969). *Trans. Met. Soc. AIME* **245**, 75.

Rice, S. A. and Allnatt, A. R. (1961). *J. Chem. Phys.* **34**, 2144.

Rice, S. A. and Gray, P. (1965). *The statistical mechanics of simple liquids.* Interscience Publishers, New York.

Rice, S. A. and Kirkwood, J. G. (1959). *J. Chem. Phys.* **31**, 901.

Rice, S. A. and Nachtrieb, N. H. (1967). *Adv. Phys.* **16**, 351.

Richardson, F. D. (1974). *Physical chemistry of melts in metallurgy.* Academic Press, London.

Rohlin, J. and Lodding, A. (1962). *Z. Naturforsch.* **17a**, 1081.

Roll, A., Felger, H., and Motz, H. (1956). *Z. Metallk.* **47**, 707.

Roll, A. and Motz, H. (1957). *Z. Metallk.* **48**, 272.

Roscoe, R. (1958). *Proc. Phys. Soc.* **72**, 576.

Roscoe, R. and Bainbridge, W. (1958). *Proc. Phys. Soc.* **72**, 585.

Rothman, S. J. and Hall, L. D. (1956). *Trans. AIME* **206**, 199.

Rothwell, E. (1961–2). *J. Inst. Metals* **90**, 389.

Sacris, E. M. and Parlee, N. A. D. (1970). *Met. Trans.* **1**, 3377.

Saito, T. and Sakuma, Y. (1967). *J. Japan Inst. Metals* **31**, 1140.

Saito, T., Shiraishi, Y., and Sakuma, Y. (1969). *Trans. ISIJ.* **9**, 118.

Samarin, A. M. (1962). *JISI.* **200**, 95.

Scatchard, G. (1937). *Trans. Faraday Soc.* **33**, 160.

Schenck, H., Frohberg, M. G., and Hoffman, K. (1963). *Arch. Eisenhüttenw.* **34**, 93.

Schrödinger, E. (1915). *Ann. d. Phys.* **46**, 413.

Schytil, F. (1949). *Z. Naturforsch.* **4**, 191.

Seydel, U. and Kitzel, W. (1979). *J. Phys. F* **9**, L153.

Shapiro, J. N. (1970). *Phys. Rev.* **B1**, 3982.

Shimoji, M. (1967). *Adv. Phys.* **16**, 705.

Shimoji, M. (1977*a*). *Liquid metals: an introduction to the physics and chemistry of metals in the liquid state.* Academic Press, p. 222 (after Kitajima, M. (1976). Thesis Hokkaido University).

—— (1977*b*). —— p. 231.

—— (1977*c*). —— p. 227.

Shiraishi, Y. and Tsu, Y. (1982). The 140th Committee, the Japan Society for the Promotion of Science (JSPS), Rep. No. 129, Dec. 1982.

Singh, A. K. and Sharma, P. K. (1968). *Can. J. Phys.* **46**, 1677.

Skapski, A. S. (1948). *J. Chem. Phys.* **16**, 389.

Sokolov, L. N., Katz, Ya L., and Okorokov, T. N. (1977) 4. *Izv. Akad. Nauk SSSR, Metally*, p. 62.

Solar, M. Y. and Guthrie, R. I. L. (1972). *Met. Trans.* **3**, 2007.

Spells, K. E. (1936). *Proc. Phys. Soc.* **B48**, 299.

Steeb, S. and Bek, R. (1976). *Z. Naturforsch.* **31a**, 1348.

Steinberg, D. J. (1974). *Met. Trans.* **5**, 1341.

Stratton, R. (1953). *Phil. Mag.* **44**, 1236.

Sugden, S. (1922). *J. Chem. Soc.* **121**, 858.

Sugden, S. (1924). *J. Chem. Soc.* **125**, 27.

Sutherland, W. (1905). *Phil. Mag.* **9**, 781.

Suzuki, K. and Mori, K. (1971). Tetsu-to-Hagané **57**, 2219.

Swalin, R. A. (1959). *Acta Met.* **7**, 736.

Swalin, R. A. and Leak, V. G. (1965). *Acta Met.* **13**, 471.

Takeuchi, S. and Endo, H. (1962*a*). *J. Japan Inst. Metals* **26**, 498.

Takeuchi, S. and Endo, H. (1962*b*). *Trans. JIM* **3**, 30, 35.

Takeuchi, S. and Iida, T. (1972). Quoted by Kihara, H., Okamoto, I., and Iida, T., *Trans. JWRI. (Japan)* **1**(1), 33.

Takeuchi, S. and Misawa, M. (1971). *Bussei Kenkyu* **16**(5), 654.

Takeuchi, S., Morita, Z., and Iida, T. (1971). *J. Japan Inst Metals* **35**, 218.

Tamaki, S., Ishiguro, T., and Takeda, S. (1982). *J. Phys. F* **12**, 1613.

Tamamushi, B. *et al.* (eds.) (1981). *Iwanami dictionary of physics and chemistry (Iwanami rikagaku jiten)* (3rd edn, enlarged). Iwanami Shoten, Publishers, Tokyo.

Taylor, J. W. (1954). *Metallurgia* **50**, 161.

Taylor, J. W. (1954–5). *J. Inst. Metals* **83**, 143.

Taylor, J. W. (1955). *Phil. Mag.* **46**, 867.

Thresh, H. R. (1962). *Trans. AMS* **55**, 790.

Thresh, H. R. (1965). *Trans. Met. Soc. AIME* **233**, 79.

Thresh, H. R. and Crawley, A. F. (1970). *Met. Trans.* **1**, 1531.

Thresh, H. R., Crawley, A. F. and White, D. W. G. (1968). *Trans. Met. Soc. AIME* **242**, 819.

Tomlinson, J. L. and Lichter, B. D. (1969). *Trans. Met. Soc. AIME* **245**, 2261.

Toxvaerd, S. (1971). *J. Chem. Phys.* **55**, 3116.

Toye, T. C. and Jones, E. R. (1958). *Proc. Phys. Soc.* **71**, 88.

Tsu, Y., Shiraishi, Y., Takano, K., and Watanabe, S. (1979). *J. Japan Inst. Metals* **43**, 439.

Tsu, Y., Suenaga, H., Takano, K., and Shiraishi, Y. (1982). *Trans. JIM* **23**, 1.

Tsu, Y., Takano, K., and Shiraishi, Y. (1985). *Bull. Res. Inst. Mineral Dressing Met.*, Tohoku Univ., **41**, 1.

Turkdogan, E. T. (1980). *Physical chemistry of high temperature technology.* P. 88, Academic Press.

Ubbelohde, A. R. (1965). *Melting and crystal structure.* 8 p. 170, Clarendon Press, Oxford.

Vadovic, C. T. and Colver, C. P. (1970). *Phil. Mag.* **21**, 971.

Veazey, S. D. and Roe, W. C. (1972). *J. Mater. Sci.* **7**, 445.

Vertman, A. A. and Samarin, A. M. (1960). *Dokl. Akad. Nauk. SSSR.* **132**, 572.

Vertman, A. A. and Samarin, A. M. (1969). *Metody issledovaniya svoisty metallicheskikh rasplavov,* p. 28. Nauka, Moskva.

Vertman, A. A., Samarin, A. M., and Filippov, E. S. (1964). *Dokl. Akad. Nauk SSSR.* **155**, 323.

Viswanath, D. S. and Mathur, B. C. (1972). *Met. Trans.* **3**, 1769.

Wagner, C. N. J. (1982). *Liquid metals: chemistry and physics* (ed. S. Z. Beer). Marcel Dekker, New York, p. 286.

Wainwright, T. and Alder, B. J. (1958). *Nuovo Cimento* **9**, Suppl., 116.

Walls, H. A. and Upthegrove, W. R. (1964). *Acta Met.* **12**, 461.

Wanibe, Y., Takagi, T., and Sakao, H. (1975). *Arch. Eisenhüttenw.* **46**, 561.

Waseda, Y. (1980a). *The structure of non-crystalline materials: liquids and amorphous solids.* P. 48, McGraw-Hill, New York.

—— (1980b). —— Ch. 8.

—— (1980c). —— p. 54.

—— (1980d). —— Appendices 8, 9.

—— (1980e). —— p. 72.

—— (1980f). —— p. 198.

—— (1980g). —— p. 203.

Waseda, Y. and Ohtani, M. (1973). *Sci. Rep. Res. Inst. Tohoku Univ.* **24A**, 218.

Waseda, Y. and Ohtani, M. (1975). Tetsu-to-Hagané **61**, 46.

Waseda, Y. and Suzuki, K. (1972). *Phys. Status Solidi.* **b49**, 643.

Waseda, Y., Suzuki, K., and Takeuchi, S. (1969). Tetsu-to-Hagané **55**, S444.

Waser, J., and Pauling, L. (1950). *J. Chem. Phys.* **18**, 747.

Watanabe, S. and Saito, T. (1968). *Bull. Res. Inst. Mineral Dressing Met.* Tohoku Univ. **24**, 77.

Wen-Po, W. (1937). Phil. Mag. **23** 33.

White, D. W. G. (1962). *Trans. ASM* **55**, 757.

White, D. W. G. (1966). *Trans. Met. Soc. AIME* **236**, 796.

White, D. W. G. (1968). *Met. Rev.* **13** Rev. 124, 73.

White, D. W. G. (1972). *Met. Trans.* **3**, 1933.

Williams, D. D. and Miller, R. R. (1950). *J. Am. Chem. Soc.* **72**, 3821.

Wilson, J. R. (1965a). *Met. Rev.* **10**, 542.

—— (1965b). —— 558.

—— (1965c). —— 572.

—— (1965d). —— 573.

—— (1965e). —— 454.

—— (1965f). —— 553.

Wittenberg, L. J. and DeWitt, R. (1972). *J. Chem. Phys.* **56**, 4526.

Wittenberg, L. J. and DeWitt, R. (1973). *The properties of liquid metals* (ed. S. Takeuchi). P. 555, Taylor and Francis, London. (Proc. 2nd Int. Conf.).

Wood, W. W. and Jacobson, J. D. (1957). *J. Chem. Phys.* **27**, 1207.

Woodcock, L. V. (1976). *J. Chem. Soc. Faraday Trans. II* **72**, 1667.

Woodward, J. G. (1953). *J. Acoust. Soc. Am.* **25**, 147.
Worthington, A. M. (1885). *Phil. Mag.* **20**, 51.
Yang, L., Kado, S., and Derge, G. (1958). *Trans. Met. Soc. AIME* **212**, 628.
Yao, T. P. (1956). *Giesserei Techn.-Wiss. Beih.* **16**, 837.
Yao, T. P. and Kondic, V. (1952–3). *J. Inst. Metals* **81**, 17.
Young, D. A. and Alder, B. J. (1971). *Phys. Rev.* **3A**, 364.
Ziebland, H. (1969). *Thermal conductivity* (ed. R. P. Tye). Vol. 2, Ch. 2, Academic Press, London.
Ziman, J. M. (1961). *Phil Mag.* **6**, 1013.
Ziman, J. M. (1964). *Adv. Phys.* **13**, 89.

SUBJECT INDEX

Page references in italics are to tables, figures, or footnotes.

AUTHOR INDEX

Page references in italics are to tables, figures, or footnotes.